COMPLEX STEEL STRUCTURES

NON-ORTHOGONAL GEOMETRIES IN BUILDING WITH STEEL

COMPLEX STEEL STRUCTURES

NON-ORTHOGONAL GEOMETRIES IN BUILDING WITH STEEL

TERRI MEYER BOAKE

BIRKHÄUSER
BASEL

PROJECT PROFILES

CONTENTS

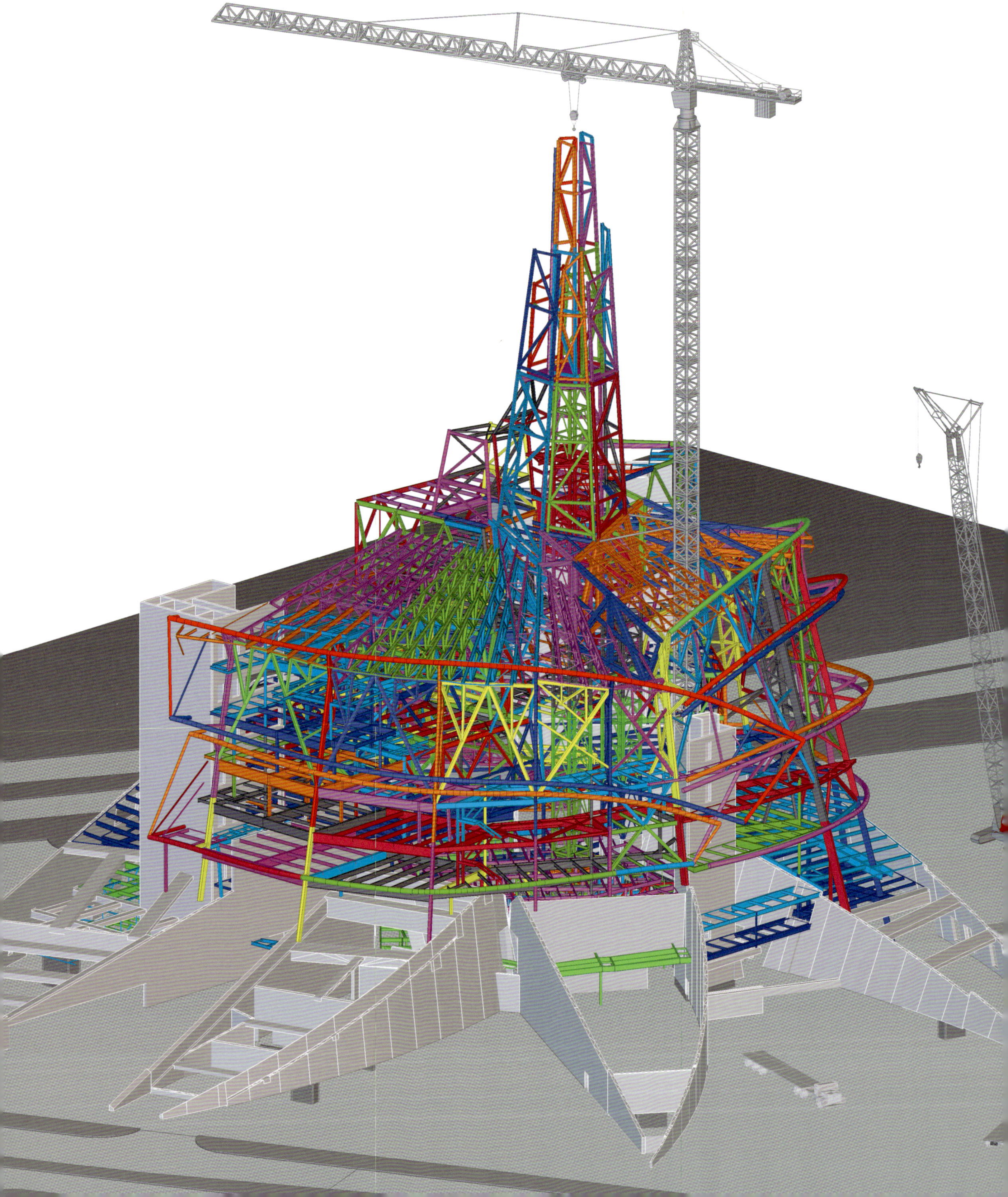

FOREWORD

Walters Group is proud to be the sponsor of *Complex Steel Structures* by Terri Meyer Boake. Walters Group is a Canadian vertically integrated steel fabricator/constructor. Our journey started in 1956 as a miscellaneous metals' fabricator. Today Walters Group completes projects across North America, ranging from public art installations to major industrial facilities, to iconic infrastructure and cultural landmarks such as museums, feature lobbies and atriums, sports facilities, educational facilities, performance spaces and airports.

Our relationship with Professor Meyer Boake stretches back years. We have been the beneficiary of her insight into the world of architecture, and especially about the way that architecture relates to steel. Terri has been a frequent visitor to many of our (and other) construction sites. I have witnessed first-hand that she is not afraid to get close to the action, boots and harness on and camera at the ready, whether at grade or on the steel hundreds of meters in the air. She brings an intense curiosity to her work, and I believe she has made a difference in the evolution of steel as a construction material.

Technology has facilitated the proliferation of complex structures globally; however, it is important to note that it hasn't changed mankind's fundamental ability to build complex structures. Rather, technology has dramatically improved the time and cost certainty of doing so. Complex structures are, by their very nature, challenging. Simply applying the latest software technology is insufficient to ensure a successful outcome. Paramount to success is the realization that complex structures generally harbor complex problems. Only by thoroughly understanding and mitigating these challenges early in the execution process can the probability of success be elevated.

Steel is a stable and predictable material, eminently workable, sustainable, safe and durable. As the boundaries of architecture continue to expand globally, steel will remain a material of choice. Equally, as the architecture, construction and engineering industries continue their march towards greater modularity and parametric (BIM) modeling, the strength, stability and versatility of steel will continue to not only provide the supporting elements of structures, but also substantially contribute to the architectural language.

We believe that this book is a valuable resource to those delving into the world of steel construction and particularly Complex Steel Structures. I would like to thank Professor Meyer Boake for helping showcase wonderful steel structures globally, and for providing students of steel design with such an exceptional learning resource.

Walter Koppelaar
Chairman, Walters Group

This Tekla Structures image, prepared by Walters Group, of the Canadian Museum for Human Rights shows the complex challenges when bringing projects to a quality-built resolution. A high level of innovation by the steel fabricators and erectors is required to bring such ideas to life. This museum employs a concealed diagrid structure to support a massive stone façade; Architecturally Exposed Structural Steel on the interior to create the aesthetics desired to embody human suffering in material form; exposed galvanized steel to form the Tower of Hope; finer AESS to provide wind bracing for the sloped glazed façade; a large curved tubular rail to support the glazing system; and major concealed nodal elements to handle areas of extreme load transfer. Projects such as these present a challenge of the highest order, one that it is hoped will be aided by the level of information provided in this book.

ACKNOWLEDGMENTS

This publication has been made possible through the generous support of Walters Group, a first-class Canadian steel fabricator with one of the most innovative teams with whom I have had the pleasure to work. Walters Group is a very family- and people-focused business that goes the extra distance to support those in need in their community. Safety and excellence are a top priority. You have taught me so much.

Thank you as well to my editor Andreas Müller for his expertise, input and support and to Reinhard Steger and his team for their loving attention to the graphic layout of this book, and the three others that precede this work. This book has its place in a long journey through steel design, beginning with *Understanding Steel Design: An Architectural Design Manual*, published by Birkhäuser in 2011, *Diagrid Structures: Systems, Connections, Details*, published in 2013, and *Architecturally Exposed Structural Steel: Specifications, Connections, Details* in 2015. *Complex Steel Structures: Non-Orthogonal Geometries in Building with Steel* emerged as a critical crossover between my research on Architecturally Exposed Structural Steel and diagrid structures.

Most of the images used in the book are of my taking. I am thankful to have been able to procure some key construction images from photographers. Process shots are critical to developing a more complete understanding of this level of complexity.

Although the research and writing of this work commenced in 2016, it suffered a considerable delay as a result of serious health issues in my family that are thankfully now resolved. It gave me pause to think seriously about the relative importance of architecture when health-based emergencies arise. However, even in times of crisis, the spaces we inhabit have the ability to raise our spirits and give us hope. The human spirit still needs dreams and inspiration. The completion of this book is taking place while all involved are in one sort or another of self quarantine as a result of COVID-19. Sitting at my computer selecting images allowed me to appreciate the importance to my work of the ability to travel. My thoughts and ideas focus on first-person research as I engage with architectural works and active construction sites. I long for a time in the future when this will again be possible.

Lastly I wish to thank my family for the continued support of and pride in my publication endeavors. None of this would be possible without my team.

The parametrically driven project of the Phoenix New Media Center in Beijing, China, 2014, designed by BIAD, challenged all aspects of the design, fabrication and erection process. The smooth curved lines required all-welded splice connections to support the aesthetic inspirations of the project.

PREFACE

This book builds strongly on materials addressed in my three previous books. These works will be referenced throughout.

For the most part, complexity in architecture is aesthetically driven. In that complexity is a goal in most cases, rather than a requirement. There are multiple paths that can be taken and methods used to achieve more or less the same results. This book will look at a range of strategies towards achieving complexity in steel, be it visually or configurationally motivated. Two important terms that have been derived from best practices discussed in *Architecturally Exposed Structural Steel* will factor into decision making: the *distance factor* and the *multiplication factor.* The distance to view suggests that fabrication detailing can be softened when the steel is not viewable up close. The multiplication factor acknowledges that costs increase when more expensive details are required on dozens to hundreds of connections, and that less expensive detailing methods might be a prudent choice.

This book is divided into three distinct parts.

PART ONE

The Evolution of Complex Geometry takes us through the emergence of complexity in steel design and the push and pull in the desire to create more complex architecture, responding to the growing ability to design and construct as a direct result of advances in software. Much of the initial approach to steel detailing was born in the High Tech period, focusing on modularity and visual complexity rather than configurational complexity. Deconstructivism and parametric design have shifted architecture towards geometries that exclude modularity and repetition. This resulted in a configurational complexity that increased the challenges of fabrication and erection. *Part One* introduces some definitions and approaches that can be used to evaluate and plan for different types of complexity.

PART TWO

Designing for Complexity – Practical Considerations examines the fundamental issues associated with the realization of configurationally complex steel projects. The pertaining fabrication methods require a much higher level of communication among the team members, which now routinely includes the fabricator as a consultant at early stages of design. Finishes become more challenging as we work with increased exposure of the steel – interior situations addressing fire protection and exterior situations working to combat weather and corrosion. Construction processes can be extremely challenging when asymmetrical lifts work against gravity. The erection process, limited staging areas and splice considerations need to inform the overall approach to design and detailing, as they impact the overall aesthetic drive of the project.

PART THREE

Implementation Strategies examines in detail a wide range of projects loosely classified according to their respective dominant approach and type of complexity, as set out in the first part of the book. The project profiles are not intended to be comprehensive. Naturally, each project may include more than one attribute; the focus will be placed on the aspect that can provide optimal information to assist in your own decision-making process. The idea behind this part of the book is to examine a range of approaches that can be adopted or adapted to inform the design process.

Designers need the tools to appreciate the means and methods to design, fabricate and build with complex steel.

THE RISE OF COMPLEXITY

CHAPTER 1

In order to understand how we can best undertake steel design for complexity in the present Post-Fordist age of Parametricism, it is necessary to trace its evolution. Many of the current techniques for managing the design and fabrication of complex steel systems are closely based on practices that emerged from Modernism throughout the High Tech period. These strategies were subsequently modified and adapted over time. Current best practices still need to acknowledge the craft-based work of the ironworker that works in parallel to increasingly digitally driven practices in design and fabrication. Many present approaches to complex steel structures have been derived from these emerging trends.

1. THE MODERN CONDITION: STARTING POINT

The Modern buildings that predated the emergence of complex steel structures tended towards the use of a fairly regular column grid with a clear hierarchy of spanning systems exhibiting consistent characteristics: corrugated decking on joists or beams spaced at regular intervals, sitting on beams or girders. Larger spans were often accommodated with trusses, initially planar and eventually spatial to manage larger spans. This predisposition towards geometric simplicity aligned well with limitations on engineering practices of the period, practices that preferred to resolve structures into two-dimensional, determinate force systems. This method worked with the dominance of pin-and-hinge-type connections between members that could be resolved into simple vertical and horizontal force systems. Span and sizing tables became the backbone of standard structural engineering. The majority of calculations were made using a slide rule until the advent of personal calculators in the early 1970s.

The detailing of the steel in the La Villette project by Bernard Tschumi in Paris, France, 1987, reflects well the transition to a new type of steel detailing in Deconstructivism. The force-differentiated structural system that had been developed in High Tech architecture is transformed to stabilize the eccentric loading and twisting shapes of the new style. The work includes angled as well as curved elements.

Stylistically the overall tendency during the Modern period was towards concealment of the structural steel system. The detailing was left to the discretion of the structural engineer to develop the most efficient solution. This led to conservative detailing directed at minimizing the weight and thereby the cost of steel. Steel buildings were generally priced by their tonnage, as the complexity of the connection details was in the majority of cases not relevant. When the structure was expressed, the detailing and regularity tended to become clearer, cleaner and more regular. When exposed, connections tended to be welded so as to suppress the technical aspects of the joinery. This fed well into the Modern aesthetic that was more formally driven and eschewed technical details.

The Farnsworth House in Plano, IL, USA, designed by Ludwig Mies van der Rohe , 1951, is characteristic of the supreme simplicity of detailing in Modern exposed steel. This project used extensive welding and remediation in the creation of connections that were extensively concealed to create an air of mystery as to the method of attachment. This was typical of Mies' work. Obvious bolted connections were avoided.

The New National Gallery in Berlin, Germany, designed by Ludwig Mies van der Rohe, 1968, used an exposed steel structural system. The avoidance of bolted connections creates a sculptural aesthetic to the steel.

The detailing of the gallery made predominant use of welding to visually suppress the steel to steel connections, both within and between the fabricated members. Modified standard structural steel (wide-flange or Universal) sections were used throughout the project. The geometries allowed for repetitive detailing.

2. A CHANGE OF STYLE

There are usually reasons for stylistic changes, particularly when they are so extreme. There are also significant reasons for resistance to change. The past 50 years have witnessed remarkable changes in the form of architecture and its structural systems. From the exquisite formal simplicity of the Modern Movement emerged the exuberant exposed structural detailing of High Tech architecture and the chaotic appearance of Deconstructivism. These trends have evolved into more fine-tuned structural solutions suited to the current state of Architecturally Exposed Structural Steel. Parametric design is further challenging steel fabrication and erection strategies.

Efficiency and accessibility is necessary for an idea to become mainstream. Complexity in architecture is dependent on advances in computing power. The ability to execute the design must be present in order to bring advanced thoughts to the point of constructable reality. This is similar to the visionary work of Étienne-Louis Boullée in the 1700s. His immense ideas for the Bibliothèque Nationale and Metropole were unconstructable using even their most advanced stereometric applications in stone. Had such ideas been proposed after the invention of reinforced concrete, as were the innovative and complex structures of Pier Luigi Nervi, realization could have been possible. 3D printing is currently being heralded as the answer to the construction of complex geometries, including those in steel. However, it too is ahead of its time in terms of accessibility, cost and the scale and materiality of what can be reasonably produced.

Over time the detailed design and fabrication of steel structures has had to respond to the desire for stylistic changes. The positive and progressive response of the steel fabrication industry also might be seen to have permitted many aesthetically motivated changes to happen and flourish. The alignment between the present capabilities of design, fabrication and erection techniques is quite closely matched to the aesthetic and finish expectations of the majority of Architecturally Exposed Structural Steel (AESS) projects. Yet at the present time there is a growing gap between the far-reaching digital desires of parametric design and the continuing craft-based realities of a significant portion of steel fabrication and erection processes.

The architectural definition of complexity in Robert Venturi's 1966 book *Complexity and Contradiction in Architecture*, although written well before the onset of Deconstructivism and subsequently, Parametricism, may well be seen to apply:

"Architects can no longer afford to be intimidated by the puritanically moral language of orthodox Modern architecture. I like elements which are hybrid rather than 'pure,' compromising rather than 'clean,' distorted rather than 'straightforward,' ambiguous rather than 'articulated,' perverse as well as impersonal, boring as well as 'interesting,' conventional rather than 'designed,' accommodating rather than excluding, redundant rather than simple, vestigial as well as innovating, inconsistent and equivocal rather than direct and clear. I am for messy vitality over obvious unity."[1]

[1] Venturi, Robert. *Complexity and Contradiction in Architecture*. Museum of Modern Art, New York, 1966.

Theoretically the heightened level of complexity that we are currently seeing in steel structures as they support architectural design reflects the shift from Mies van der Rohe's "Less is more" toward Venturi's "Less is a bore," and from the Modernist era of mass production reflected in Fordism to the mass customization that dominates design in the Post-Fordist era. This book discusses the complexity in the design and fabrication of primarily non-orthogonal steel structural systems – as they either support the physical form of the architecture (concealed steel) or are the main focus of the architectural expression (exposed steel). The notion of exposure is a significant departure from the Venturi definition of complexity that was intended to speak more to the form and details of Postmodernism that typically placed the steel in a concealed service condition and which typically disregarded structural materials.

3. THE INTRODUCTION OF COMPLEXITY INTO ARCHITECTURAL DISCOURSE

Complexity was first introduced in the innovative steel structural systems of Buckminster Fuller and Frei Otto. The rigid modularity of Fuller's geodesic dome, constructed as the United States National Pavilion for Expo 67 in Montreal, Canada, contrasts sharply with the Olympiastadion for the Olympic Games in Munich in 1972 by Frei Otto and Günter Behnisch. Otto's stadium is referenced as being *proto-parametric* in its working method, designed and constructed in a time that predated the formal ability of parametric design enabled by computers. In the design of the tensile structure Otto used form-finding models which have been considered a form of analog material computation.

While these works were remarkable in their respective complex structural explorations, they failed to change the course of architectural design at the time. This was in part due to their unique natures and peculiarities, but more so to their lack of applicability to the majority of projects of the period. Additionally, the projects were driven by engineering, putting them immediately outside of the mainstream discourse of Modern architecture of the time. This might be cited as a pivotal point in the history of Modern architecture: a deep divide is established between the mainstream Modernists that adhered to orthogonal geometries and concealed structural steel systems, and architects that were quite comfortable with pushing the limits in structural steel design, to the point of confidence in its exposure, as was the case with High Tech architecture.

The geodesic form of the USA Pavilion for Expo 67, designed by Richard Buckminster Fuller, and the 1972 Olympic Stadium, designed by Frei Otto and Günter Behnisch, stretched the limits of steel technology. Even if failing to change the Modern architecture of the period, their approaches and detailing succeeded in providing a starting point for the High Tech movement. The rigid form of Fuller's geodesic dome forms an extreme contrast with the fluidity and adaptability of the stadium.

4. EMERGENCE OF PROCESS-BASED THINKING

Is it architecture or engineering? This seems to be a major point of discussion at the onset of High Tech architecture, one that determined whether projects or methods were included in mainstream architectural discourse. I suspect that the prior work of Fuller and Otto was classified as engineering by the theorists of the time and as such served as a point of dismissal. This seems to initially hold true also for High Tech architecture, whose structural explorations seem to have been criticized for their lack of relevance to mainstream Modern architectural discourse of the time.

An examination of the seminal theoretical texts by Frampton, Scully and others, addressing the period that covers the transition from High Modern to Postmodern, reveals a tendency to disregard and even exclude High Tech projects, although the work was underway during the same period. What is essential to the line of questioning I am exploring pertains to the process as it impacts the final product of architecture. Traditional theory tends to focus on the product. Including the process creates a radically different way of thinking through the challenges of complex steel design. This different approach to process-based thinking is clearly evident when examining the presentations that Foster + Partners give of their work. It includes not only design sketches but also views of the fabrication and construction of the projects. Their involvement in the fabrication process contributes to the high level of excellence in fabrication-related aspects of their projects.

Even though Deconstructivism garnered significant attention in theoretical writings, its process-based technical issues also remain unaddressed. A reason may be the lack of technical expertise of the writers of the time, accompanied by an ongoing disinterest in technical issues. This follows the long-standing tendency to look at the finished product rather than the process of design and construction. For normative Modern buildings the process was indeed barely interesting and fairly routine. When construction moved towards the prefabrication of highly complex elements and their assemblage on site, and subsequently to the challenge of more chaotic geometries, this should have changed.

In spite of the tendency by theorists to eschew the High Tech movement, the Sainsbury Centre for the Arts by Foster Associates was an extremely well-published project. Whereas many of the early High Tech projects were industrial or commercial in nature (Inmos Microprocessor Factory, Richard Rogers, 1982, and Renault Centre, Foster Associates, 1982), the use of the genre for an art gallery was a typological departure and piqued interest in what was perceived as a change in direction for the work of the firm. In an essay by Andrew Peckham in a special issue of the British *AD Journal Profile 19*, the building typology of Sainsbury is seen to be critical for the firm to integrate "the art of architecture" into their established systems approach to detailing. Sainsbury Centre made the cover of the American magazine *Architectural Record* in 1979, but in a special issue called "Engineering for Architecture" as opposed to the regular monthly issue. This special recognition again set the project apart. However, the completion of the Hong Kong and Shanghai Bank by Foster + Partners in 1986 was seen fit to overtake entire issues of *Progressive Architecture* and *The Architectural Review* with some of the most complete and detailed technical coverage ever in those journals.

The nature of the structure of Sainsbury Centre for the Arts was remarkably different from that of the Centre Pompidou in Paris, which was on the boards at approximately the same time. The developmental independence of these two major projects is of interest, as it allowed for the simultaneous innovation of highly differing details and approaches to expressed modular-based steel design. Both achieved a level of visual complexity that exceeded that of previous "serviced sheds,"

The Hong Kong and Shanghai Bank Headquarters in Hong Kong, designed by Foster + Partners, 1986, pushed innovative steel construction technologies to the forefront. Its unique structural system opened a new era of skyscraper design.

a phrase coined by Reyner Banham. Both exploited the new tubular steel sections and created an appropriate innovative suite of details. We see significant custom fabrication in the projects as a new design language was written. Both projects had to address the inherent instability of the light-framed "kit of parts" systems, and in response developed methods of bracing that were able to be discreetly incorporated into the overall aesthetic. The complexity of these projects tended to be constrained to the custom design of a number of elements, which were then subject to mass prefabrication, repetition and on-site assembly, with a preference for bolted and pinned connections. The geometries still tended towards the orthogonal; however, the complex detailing was a step towards less regular geometric solutions. This is particularly true for the bracing systems that in the future would be needed for eccentric geometries.

The Sainsbury Centre for the Arts in the UK, designed by Foster Associates, 1978, represents what could be called a "visual complexity," the result of an overlay of fine exposed details that lent themselves to modular fabrication.

The Centre Pompidou in Paris, France, designed by Piano and Rogers, 1977, developed a set of purpose-designed connection details that were repeated throughout the project. Its visual complexity is the result of a layering of fine bracing details over the tubular frame, as well as the introduction of exposed mechanical and conveyance systems. Overall the project keeps to an orthogonal layout.

A major shift towards genuine complexity in a High Tech project emerged in Nicholas Grimshaw's solution for Waterloo International Terminal in London, UK, that opened in 1993. It was the largest project that the firm had undertaken to date, and the aesthetic impetus was "heroic" to fit with the terminus function for the new Eurostar train. The site for the station was extremely constricted, curving alongside an existing rail corridor as well as varying in its overall width. The arches needed to range from 55m to 35m in width. It was recognized that the budget and time could not accommodate a high level of customization and so the firm looked at innovative means to respond to the geometric variety. To keep glazing costs down, regular rectangular panes were used with a changing overlap to respond to dimensional variations. Custom stainless steel attached to the primary truss castings also accommodated this variation, as they were able to rotate and lock the angle of the glass in place. The drive in the project was to create a comprehensive set of details that could be easily varied and fit to the changing shape, simplifying overall construction at the same time. It was recognized that cleaning would be an ongoing issue, so special brushes were designed to coordinate with the shape of the components. Although the terminal was closed in 2007, it was deemed of significant architectural value to be awarded a refurbishment, reopening in 2018.

An aerial view of Waterloo International Terminal, London, UK, designed by Grimshaw Architects, 1993. The curved truss structure that comprises the roof was required to change its shape along the length of the terminal due to site restrictions, forcing innovation in the design.

A custom stainless steel casting was designed for the attachment of the geometries of varying inclination of the overlapping glass panes. This invention pushed the evolution from the typical orthogonal serviced shed to structures that could respond quite nimbly to geometric changes, without much customization of the components.

The work of Santiago Calatrava runs parallel to High Tech architecture but is not formally considered to fall within this grouping. He was educated as an engineer and architect, and so was able to work with his structural systems on a unique level. His steel detailing was very different from the work of Norman Foster, Richard Rogers and Nicholas Grimshaw, particularly with the absence of lightweight tensile bracing systems upon which many of the early High Tech projects greatly relied. In Calatrava's work, stability is achieved more by the use of forms, balance and extensive moment-resisting connections.

Oriente Station in Lisbon, Portugal, designed by Santiago Calatrava, 1998, uses a variety of highly customized steel shapes.

Throughout history, engineering and fabrication requirements that fall beyond the scope of mainstream architectural education have served to push much of the potential of steel as a non-orthogonal exercise aside. And when non-orthogonal structural strategies are adopted to support oddly geometric or chaotic spaces, the architect often chooses to clad the steel. This can be part of an aesthetic desire, fireproofing requirements or cost concerns, but is often also done for lack of confidence and control in terms of steel detailing. What set the work of the High Tech architects apart was their extreme comfort in pushing steel detailing to the forefront of the design, making it the actual architectural heart of the project. In particular, later High Tech projects went beyond the simple aesthetic of the serviced shed and its use of more standardized components. This tendency reflects back to Structural Rationalism in the works of Henri Labrouste in the 19th century which was also unique and without widespread influence throughout that period of history.

5. DECONSTRUCTIVISM

The theory of deconstruction and the influence of Jacques Derrida on the Deconstructivist movement in architecture was less critical to the development of the details of complex steel structures. However, the overall outcome of his influence caused the entire tradition of structural design in terms of balance, symmetry and a general acknowledgment of the forces of gravity to be tossed aside. Orthogonal geometries gave way to non-orthogonal ones, and new methods of calculation, fabrication and erection were required.

The MoMA exhibition in New York City in 1988 on Deconstructivism[2] pushed the new style into the center of the architectural theory debate. Its deviation from Modernism and Postmodernism was clear in the works of Peter Eisenman, Wolf Prix, Rem Koolhaas, Frank Gehry, Zaha Hadid, Daniel Libeskind and Bernard Tschumi. Mark Wigley, the curator of the exhibition together with Philip Johnson, concluded that what could be said about all of them was that they employed skewed lines and irregular angles – or, in Wigley's definition, "an unstable, restless geometry, the kind that relies on the diagonal overlapping of rectilinear or trapezoidal bars." Wigley called this an "uneasy alliance" given the disparate nature of their works.

[2] Johnson, Philip and Mark Wigley. *Deconstructivist Architecture*. The Museum of Modern Art: Distributed by New York Graphic Society Books, Little Brown and Co. 1988.

Wigley cited "buildability" as one key factor separating Deconstructivists from the Russian Constructivists of the 1920s, whose art provided formal if not philosophical inspiration for these architects. Wigley felt that the built works were making critical statements about Modern architecture through extreme contrast. Coop Himmelb(l)au and Zaha Hadid were seen to question its structural and spatial conventions. Urban implications were addressed in the works of Koolhaas and Tschumi. Gehry's work addressed sculptural possibilities, while Eisenman's and Libeskind's pieces of that period looked more at connections to literature and philosophy. Deconstructivism's relationship to the developments of High Tech architecture, particularly in terms of the technical or process-based details of steel design, was not generally discussed at the time, even though there is visual evidence of a relationship in some of the detailing of Tschumi's La Villette project. While discussions were mostly theory- and form-based, buildability is one of the primary issues that ties Deconstructivist architecture to structural steel design.

Connection details from the largely Deconstructivist La Villette project show a clear reference to the proliferation of pin-connected details associated with the High Tech movement. The project also used force differentiation to allocate member sizes, making a clear contrast between tension and compression members. Stability challenges were expressed in the tensile support systems.

There is a significant shift in the nature of complexity in Deconstructivist steel design. Initially, Deconstructivism tended towards an adaptation of High Tech prefabrication and its associated detailing, as can be seen in the La Villette project, with less regularity and repetition. Some projects employed more rigid moment-resisting connections as a means to stabilize some of the eccentric loading on the structures. Others for the same purpose used a superposition of systems, which fed into the desire for visual redundancy in the structures.

A striking contrast between the highly repetitive framed approach of early High Tech architecture and the later, post-La Villette Deconstructivist style is the marked disappearance of light bracing systems. High Tech architecture, as an aspect of its visual complexity, often used clear force-differentiated systems to assist in the comprehension of the workings of the structural system, especially as "assembled" projects – in the sense that the actual construction sequencing could easily be imagined, even after the fact, due to the clarity in the relationship between the nature of the force systems and the visual language of the steel. X-bracing with tension rods or cables did not belong in the new Deconstructivist structural language. Alternate stabilization methods needed to be invented that fit better with the desired complex non-orthogonal aesthetic. An aesthetically driven style that had nothing to do with efficiency and everything with expression drove invention in steel detailing and structural systems.

Many Deconstructivist buildings were designed with juxtapositions of elements in order to challenge the traditional notions of highly symmetrical and ordered Modernist designs, even to the point of challenging the presentation of stability. Angles, spans and cantilevers were intended to make occupants and the public question stability, sometimes to the point of discomfort. The degree of deviation from simple orthogonal geometries varied quite significantly between the architects of the Deconstructivist movement, and also changed over time, as did the availability and evolution of parallel developments in computer-aided design. (See *Chapter 2: The Digital Revolution*).

Deconstructivists tended first to look towards an identifiable layering in structural systems, often to highlight redundancies, as in the work of the offset grids in Eisenman's early houses. These were not about the clarity of the structural function of the members at all, and the layered systems had not much to do with the detailing of the structural systems. Although the work of Peter Eisenman sits at the beginning of the Deconstructivist movement, his work does not play into developments in advancing steel construction, as the built work was not critical for his approach.

For the Museum of Pop Culture (formerly Experience Music Project), Seattle, WA, USA, designed by Frank Gehry, 2000, special nesting software was used to cut the curved web members of the undulating beams more efficiently.

The Guggenheim Museum in Bilbao, Spain, designed by Frank Gehry, 1997, uses a combination of expressed and concealed steel to create dynamic multi-storey atrium spaces.

The back side of the Jay Pritzker Pavilion in Chicago, IL, USA, designed by Frank Gehry, 2004, shows the eccentric combination of curves and sharp angles that became characteristic of his work. Heavier structural elements are used to create the bracing systems, in contrast to the lightweight bracing seen in High Tech architecture.

Eisenman's early work sits in great contrast to the designs of Frank Gehry, for whom materiality and technical advances were paramount and the built projects were the focus of the energies of his enterprise. Gehry expressed that he was trying to create movement in his architecture. This set up an intriguing challenge for a structural system which had long been relegated towards statics and stability. The impression of movement while at the same time expressing stability and providing a sense of confidence in the structure of architecture became the crux of designing his steel systems. This brought many other requirements into the performance and detailing of the structural systems, including significant amounts of eccentric loading that required unique detailing to achieve a sense of stability in the final forms. Gehry was one of the early adopters of the digital, and part of his legacy has been to turn software systems not intended for architectural practice into instruments that supported the design of the most challenging geometries. The curved and chaotic nature of Gehry's steel structures extends complexity also to the design and detailing of cladding systems.

6. TO EXPOSE OR NOT TO EXPOSE

At the time of the MoMA exhibition on Deconstructivism in 1988, mentioned previously, most of the architects represented there were at the early stages of their careers, several without any significant built works. The translation of ideologies that were present only in drawn or model form to a realized project necessitated the incorporation of the realities of construction of a challenging order. This challenge may be seen to have influenced the choice of whether or not to expose the steel in their projects.

The architects who began to embrace non-orthogonal steel as the structural basis for Deconstructivist projects tended to fall in two primary camps. There were those who used the capabilities of steel to create architectural spaces whereby the steel was in a service situation and was typically clad, and those who confidently chose to express the steel as the main event of the project. Some projects had hybrid qualities where the exposure and concealment were more selective, based on programmatic and fire protection requirements in addition to aesthetic concerns. The question of exposure resulted in major differences in the approach to structural system design, selection of member types and articulation of the associated details.

The work of Daniel Libeskind, although pushing non-orthogonal geometries, had not embraced exposing steel, rather suppressing the materiality of the structural systems. The Denver Art Museum, the Addition to the Royal Ontario Museum and the glass courtyard of the Jewish Museum in Berlin all employ highly chaotic or crystalline geometries and all clad the steel. Nevertheless, the works effectively advanced the strategies of the engineers and steel fabricators to respond to the very demanding structures. A cause-and-effect scenario: had the overall design not required such chaotic geometries, structural systems would not have been designed to support these and software systems development and steel fabrication detailing and erection strategies not extended. Where extensive shoring towers were used in Denver to support the erection of the massive angular extensions of the museum's crystalline shape over adjacent roads, such support was impossible due to traffic restrictions at the urban site of the Royal Ontario Museum. Here, the structural steel design had to be modified to make the steel diagrid self-supporting during erection, with virtually no shoring or temporary tension bracing possible.

The glass courtyard at the Jewish Museum in Berlin, Germany, designed by Daniel Libeskind, 2007, uses a chaotic structure to support a fairly typical skylight. The structural steel is clad, the cladding providing enclosure for services that run alongside the columns.

The steel structure of the Denver Art Museum, Denver, CO, USA, designed by Daniel Libeskind, 2006, uses gypsum board to clad the steel. This approach was also followed in his design for the Royal Ontario Museum.

The rugged nature of the detailing of the diagrid steel structure for the Addition to the Royal Ontario Museum, Toronto, ON, Canada, designed by Daniel Libeskind, 2007, is suppressed in deference to interior finishes. The pattern made by the fenestrations acknowledges the diagrid steel frame.

When complex steel is exposed to view and can be considered to constitute Architecturally Exposed Structural Steel (AESS), the fabrication requirements become demanding, particularly with respect to ensuring fit. Exposed steel is a critical aspect of the Seattle Central Library by OMA/LMN, 2004. The project was in design at a time that predated the introduction of more robust 3D steel detailing systems. Although the project's angular geometries are irregular and unsymmetrical, a significant degree of rationalization was required to make the project constructable. The clarity in the division between the orthogonal reinforced concrete center of the library proper and the seismic-resisting angular steel diamond grid is very clear.

This acknowledgment of the expressive role of the structure continued into the subsequent OMA design for the CCTV Building in Beijing, China, ultimately completed in 2013 (following a stall due to a devastating fire at the conjoined adjacent TVCC Building in 2008). Here, the angular geometries have been translated into two connected towers. The pattern of the diagrid that is expressed on the façade is a direct translation of the load-carrying requirements of the elements. There was a very close cooperation between the architects and the structural engineers, Magnusson Klemencic Associates (MKA) for the library and Arup for CCTV. MKA had also done the engineering for the Experience Music Project by Frank Gehry in Seattle in 2000. This reflects back to the observation made by Mark Wigley in 1988 regarding the challenge of buildability in Deconstructivist projects. We see the emerging trend towards a higher level of collaboration between architects and engineers as the primary means to ensure a good outcome. These new structural forms were requiring innovative approaches from both architects and engineers, creating a hierarchy in engineering firms in terms of their ability to respond to the demands of these new forms.

The highly systemized steel diamond grid of the Seattle Central Library, WA, USA, designed by OMA/LMN, 2004, has been detailed in a fashion that matched the capabilities of the digital and fabrication technologies of the time. The project took advantage of shop fabrication to maximize the components for the exposed steel seismic bracing system. Structural engineering by Magnusson Klemencic Associates.

The CCTV Building in Beijing, China, designed by OMA, 2013, is reflective of a high level of teamwork among the architectural and engineering team members. Such complexity necessitates a shift away from the engineer's role as a consultant and towards a more collaborative arrangement. Structural engineering by Arup.

7. PARAMETRIC DESIGN

Advances in the digital field have brought widespread use of parametric tools to assist in the design process. Parametric design and the associated algorithmic design are not necessarily associated with a style or movement but reference the use of computation in virtual modeling processes. At the extreme end lies Parametricism, representing a heightened use of parametric design that has become more distinctly style driven. Parametrics of any degree requires addressing heightened demands on steel design as it drives variability over uniformity in the members.

The advent of parametric design overlaps the Deconstructivist period and its evidence can be seen to emerge in the more complex works of Frank Gehry. More recently, it can be characterized by the works of the offices of Coop Himmelb(l)au and Zaha Hadid, continuing under the direction of Patrik Schumacher. While there is a fairly close match between the aspirations of later Deconstructivist architecture and steel fabrication/erection technologies and capabilities, there is a gap between what can be imagined in a parametric architectural model and what can economically be fabricated using the present combination of computer-assisted plus craft-based methods. Some of this gap arises from the shift from the hard angular lines of Deconstructivist architecture towards more sensuous curves. Current fabrication methods are more adept at handling angular geometries than irregular curvature. Other challenges arise from the sheer level of complexity, combined with the extreme scale of some of the projects.

Where the early explorations in the innovative steel design of Otto and Buckminster Fuller could be assessed as "engineering-driven" and were thereby dismissed from mainstream architectural discourse, parametric design, which has become very much mainstream, is architecturally driven but "engineering-reliant." It requires the highest level of collaboration among the architect, engineer and steel fabricator/erector.

The work of Coop Himmelb(l)au from the 1980s onward reflected a modification of established High Tech design. This practice was confident in their steel exposure, and the detailing suited the chaotic and angular nature of the style. More recently, many of their projects have addressed curvature, in combination with highly eccentric and asymmetrical designs placing higher demands on steel fabrication and erection.

The Rooftop Remodeling at Falkestraße, Vienna, Austria, in 1998 demonstrates some of the earliest Deconstructivist use of steel by Coop Himmelb(l)au. There is complexity and exposure but still a simplicity in detailing that recognized fabrication limitations of the time.

The twisting diagrid lattice of the BMW Welt Museum in Munich, Germany, designed by Coop Himmelb(l)au, 2007, shows mastery of the digitally integrated design of AESS and architectural space. This fully welded structure required an extreme level of precise fit and tight tolerances.

The Dalian International Conference Center, China, designed by Coop Himmelb(l)au, shown here under construction in 2011, is indicative of the exploration of the interface between advanced digital design and steel construction.

Zaha Hadid did not have a major built work until five years after the MoMA exhibition of 1988. Much of the current challenge of this practice has been in the realization of the parametrically driven work. It is with the highest levels of Parametricism that we see the biggest gap between the precision of the computer and the challenges of most constructed projects. Some of the undulating lines have tended to be constructed in reinforced concrete, as its fluidity in final form and methods of construction can more closely match the continuity of lines in the design intentions. Where the projects use steel as a support system, concealed or exposed, more careful attention has to be paid to fit the design of the steel to the cladding system it supports. It is ultimately the cladding that is burdened with the task of the curvilinear form, and to achieve true curvature in the façade the tolerances must be extremely tight and the material selection appropriate.

The Guangzhou Opera House, China, designed by Zaha Hadid, 2010, demonstrates the challenge of creating a curved stone façade to fit an asymmetrical, parametrically driven structural steel frame.

The interior of the Guangzhou Opera House shows the exposure of the steel frame. The dimensions of the triangulated cladding system are much smaller than the structural grid in order to assist in better fitting the impression of curvature. The large-scale lattice was constructed from custom box sections with precisely finished square corners. All of the connections assembled on site were fully welded and remediated to provide a seamless appearance.

Much of the challenge of parametric design lies with the wide variation in craft-based skills in the global industry. A higher than normal degree of on-site welding is required to create continuous lines. This in turn requires highly skilled site labor as well as highly qualified shop labor. There are inferences in terms of project costs associated with additional support, scaffolding and facilities required for quality on-site work. This will be discussed in detail in *Chapter 6: Erection Logistics.*

To bring current high levels of parametrically driven design aspirations into built form raises the question, "How much do architects really need to know?" when it comes to an understanding of steel fabrication detailing and construction sequence strategies. Historically, this has not been part of the architect's concern. I maintain that it now is.

And this is the subject of this book.

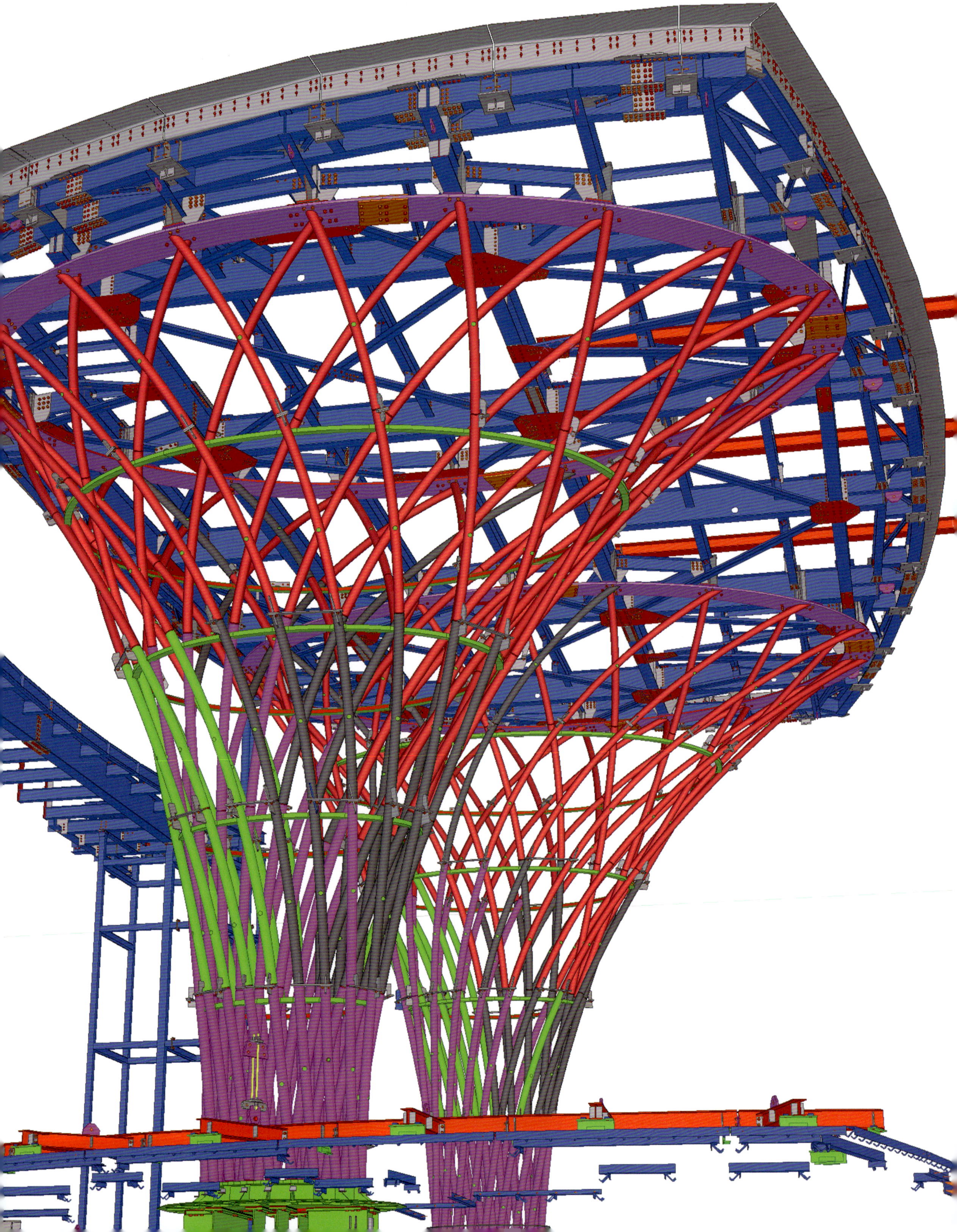

THE DIGITAL REVOLUTION

CHAPTER 2

If the ultimate goal is to build, there needs to be a close alignment between the ability to imagine a structure, the ability to calculate the structure and the tools to fabricate and erect the structure. When digital capabilities step beyond techniques that remain predominantly craft-based, we need to come up with innovative approaches to overcome the mismatch.

1. THE LINK BETWEEN COMPUTING AND COMPLEXITY

There is an intrinsic connection between advances in computing technology and the evolution of complexity in architectural design. This includes all facets of design, drawing, calculation and production. It is logical, then, that the increase in complexity of steel structural systems has followed along quite tightly with the evolution of computing systems over the last five decades. Pre-computational methods of drawing and calculation typically required the simplification of structural systems to 2D scenarios, which tended to prefer orthogonal geometries, as they were easier to calculate, draw and construct. Architects, too, tended to eschew complex geometries in their designs, as they were difficult to draw.

There has not always been a precise match between what architects can conceive, desire and draw, what engineers can reasonably calculate and what steel fabricators can economically and precisely fabricate and erect. There tend to be pushes and lags. And at present the fabrication and erection methods for steel tend to sit behind "the digital work done at the desk," for very practical reasons related to tolerance, fabrication processes, irregularities of material and human error.

The Tekla Structures image of the Brookfield Place Pavilion in New York City, NY, USA, designed by Pelli Clarke Pelli, 2013, shows the level of drawing and detailing complexity that the steel fabricator Walters Group used to guide the fabrication and erection processes.

2. HIGH TECH ARCHITECTURE SETS THE STAGE

Some of the initial design changes that predate the widespread incorporation of computing in design began to generate a need for that assistance. Some of this need might be seen as subtle, thinking of the contrast in the nature of the steel structures of the Modern Movement (Mies) to those of the High Tech period (Foster, Rogers, Piano, Grimshaw). This increase in complexity was initially characterized by the introduction of force-differentiated structural systems – those making widespread use of the exposure of lightweight tensile members to provide stability to the largely hinge-and-pin-connected prefabricated and often modular systems – as well as the forefronting and exposure of the steel systems and connection details.

The High Tech projects of the 1970s and 1980s might still have been drawn and calculated by the same analog methods as were the orthogonal Modern projects; however, the approach to design took on a set of assemblies that appeared more complex. High Tech architecture was characterized by modularity and repeated elements creating visual complexity. There was more work entailed in the calculations of the forces in the more articulated assemblies, but once done, these assemblies were mass-fabricated with the best available techniques of the period. This would include the manual transference of dimensions and angles to the steel as well as reliance on hand-crafted jigs in the fabrication shop to ensure that all like elements were fabricated to acceptable, tighter-than-normal tolerances. Mentally, this visual complexity set the stage for an acceptance of the increase in configurational complexity that was initiated by Deconstructivism.

If we look at the details of the Sainsbury Centre for the Arts by Foster Associates (1977) and the Centre Pompidou by Piano and Rogers (1976) we see quite elaborate steel detailing that has been achieved through approximately the same methods that were available and used on typical Modernist exposed steel projects. What these projects all had in common in their engineering was the ability to translate the force systems into two-dimensional scenarios for calculations. They also relied on conventional 2D orthographic shop drawings. However, it was necessary to develop completely new interpretations of standard steel connections to elevate the steel connection detailing of High Tech architecture to industrial design standards. Although occurring in advance of the onset of digitally assisted shop fabrication processes, this succeeded in tooling up the fabrication industry and also began to create a hierarchy among steel fabricators and erectors – between those who could handle more elaborate expressed steel projects and those who could not. As far as architects went, this style of architecture remained inaccessible to most due to its intrinsic dependence on technical knowledge pertaining to structural detailing and fabrication methods. It was not taught in schools or valued by the profession at large.

The custom triangular trusses forming the pinned arches that set the form for the Sainsbury Centre for the Arts in the UK, designed by Foster Associates, 1977, were repeated on a consistent module for the entire length of the building. This repetition created economies in all aspects of design, fabrication and erection.

The use of lightweight X-bracing systems was essential in the stabilization of the modular structural system for the Centre Pompidou in Paris, France, designed by Piano and Rogers, 1976. Visual complexity was a by-product of this added layer.

3. A TIMELINE OF SOFTWARE ADVANCES

1975 K. Versprille's PhD thesis "Computer-Aided Design Applications of the B-Spline Approximation Form" develops the mathematical representation of arbitrary curves suitable for computation
1977 The Apple // is the first personal computer to include graphics
1977 CADAM, the first commercial 2D CAD package, is released
1981 CATIA, one of the first 3D CAD packages, is developed, using constructive solid geometry
1982 AutoCAD 1.0 is released, using only wireframe representation
1988 CATIA is selected as CAD package for Boeing 777
1988 Frank Gehry uses CATIA Version 3 to design a 50m-long fish for the 1992 Olympic Games in Barcelona, Spain
1989 Adobe Photoshop is released
1991 form·Z is released as a CAD tool developed by AutoDesSys for all design fields that deal with the articulation of 3D spaces and forms
1993 Xsteel Version 1.0 is released
1994 Autodesk releases a 3D version of AutoCAD
1998 Rhinoceros Version 1.0 is released
1998 Maya animation software is first released
2000 Sketchup, the first web-based 3D CAD system, is released
2000 Autodesk Revit is released
2002 Digital Project is a CAD application based on CATIA V5 and developed by Gehry Technologies and Dassault Systems
2004 Tekla Structures is launched (rebranding of Xsteel)
2008 Grasshopper plug-in is added for Rhinoceros
2011 Tekla BIMsight, an open collaboration tool for construction, is launched
2013 Autodesk Revit rolls functionality of its separate software lines into one product
To the present – Rollout of cloud computing allowing high-level interoperability between systems and users

The more widespread adoption of 2D CAD software started in the late 1980s. However, these early tools mimicked hand-drawn orthographic practices in their production of standard drawing sets. These planar views gave way to 3D views as the software became more accessible, yet computers of the time were incapable of processing highly complex geometries, in particular curves. Curved lines were still comprised of faceted, straight line segments, which overpowered the capability of most office computers. So although digital drawing was to become widely adopted during the 1980s, it was not able to be a driver to change the form of architecture, only to make the contemporary practice more efficient. Offices had to contend with generational lags in the workforce as well as the high cost of kitting out the practice. It also took time for education to embrace the potential of computer-aided design.

The curved cuts on the "wishbone" elements at Pearson International Airport, 2000, predated the ability of computers to create smooth curves. The curved lines in the software of the time produced faceted shapes that required the fabricator to manipulate by hand in the shop to create smooth curves.

The value of 3D modeling was recognized early on for its ability to increase the level of flexibility and accuracy in design. The idea of this simulation assisted in the vetting of ideas through digital modeling, without much associated risk in the way. However, due to the high learning curve associated with these early specialized software packages, their adoption was not widespread in the architecture and engineering industry. Much of the software that eventually was to enable complexity in architectural design had its beginnings in the automotive and aviation industries. These specialized software systems such as Maya, which was specifically produced for the animation industry, and CATIA, which catered to the aviation and automotive industries, allowed only technically skilled architects to design 3D models with an unprecedented degree of freedom in terms of geometry and with increasingly fast processing times. This included curvature. Analog drawing methods that relied on circle templates and the reach of the compass acted as a disincentive to include much in the way even of regular curves in projects. Irregular curves required the use of French curves, which offered no mathematical means to translate the results to the fabrication shop or site. By contrast, architecture-specific programs like form·Z and later Rhinoceros met with an appetite as the user interface was much more suited to the building industry. We then saw a major shift in aesthetics that became digitally driven to the extent that drawing became unencumbered and freed from its 2D roots. Computing power increased substantially, particularly with reference to the inclusion of curvature.

4. SMOOTH CURVES

Smooth curves in graphics software were not possible until the incorporation of Bézier curves, named after Pierre Bézier, who used them in the 1960s for designing curves for the bodywork of Renault cars. Bézier curves cannot be as smooth as the subsequent B-spline function, a combination of flexible bands that pass through control points and create smooth curves. These functions enable the creation of complex shapes and surfaces using a number of points. B-spline functions and Bézier functions are applied extensively in shape optimization methods in parametric software. Even before the advent of 3D parametric modeling in architecture, Bézier and B-spline functions were enabling smooth curves in illustration and font software. In particular, the geometry possible in Rhinoceros is based on the NURBS (Non-Uniform Rational Basis Spline) mathematical model, which enables the precise representation of curves and free-form surfaces. Previous 3D modeling software had used a triangulated mesh to approximate curvature.

Much of the so-called "blob" architecture that followed the advent of 3D modeling with smooth curves put stresses on the construction industry as a result of the many anomalies of form that arose. The lack of interoperability between the software packages used by architects, engineers and fabricators was not helpful. Each member of the team would need to create their own unique digital drawing set – and so changes were not easily incorporated and there was no consistency in form, details and placement.

Although some of these new software packages were also designed to aid fabrication processes, their use was not to become widespread until the release of Tekla Structures around 2005. At the time of writing, designers' requests for accurate curvature often exceed the capabilities of the steel industry to supply the same. As explained in *Understanding Steel Design: An Architectural Design Manual,*[3]

[3] Boake, Terri Meyer. *Understanding Steel Design: An Architectural Design Manual.* Birkhäuser, 2011.

Chapter 8: Curved Steel, the methods for creating curved steel are generally not computer-driven in the same way as the design software. The predominance of pressure-driven bending processes, followed by manual curve-checking that are used to bend steel shapes, could be deemed relatively crude in comparison with the precise measurements generated by contemporary modeling software. Adaptations are required in the design, fabrication and construction process to achieve realistic and cost-effective solutions. The most accurate curved shapes can be generated in the cutting of plate steel, as the cutting mechanism is guided by software.

For the Beijing National Theatre, China, designed by Paul Andreu, 2007, the egg-shaped curved geometry was created using cut plate steel. This fabrication method is more directly computing-driven and so can ensure higher dimensional accuracy.

For the Brookfield Place Pavilion in New York City, NY, USA, designed by Pelli Clarke Pelli, 2013, bent tubular steel was used that, although requiring much variation in the bending geometry, relies on a somewhat symmetrical design aesthetic.

The Phoenix New Media Center in Beijing, China, designed by BIAD, 2013, is based on a highly parametricized geometry that needs to work with standard bending methods to create the complex curvature seen in its tubular and plate-formed curved elements.

This will be expanded by looking at several curved steel case studies in *Chapter 8: Curved Geometries*.

5. BUILDING INFORMATION MODELING

The advent of Building Information Modeling (BIM) in the early 2000s, while not adding further three-dimensional geometric capabilities to the software, did add the ability to store attributes. Where previous drawing software created lines that represented aspects of a building, BIM models are creating doors, walls, and structural systems, all with trackable material attributes. Most importantly, these systems also permit the coordination of the various systems such as structural, electrical and HVAC, which has become pivotal for the successful coordination of geometrically complex buildings.

"With BIM (Building Information Modeling) technology, one or more accurate virtual models of a building are constructed digitally. They support design through its phases, allowing better analysis and control than manual processes. When completed, these computer-generated models contain precise geometry and data needed to support the construction, fabrication, and procurement activities through which the building is realized."[4]

[4] Definition of Building Information Modeling in the Handbook of BIM, Eastman, Teicholz, Sacks & Liston, 2011.

[5] *Engineering News Record*, "Paradigm Shifting: Digital Modeling Mania Upends the Entire Building Team." Mc-Graw Hill. June 5, 2006.

There were many essential advantages to the use of Tekla Structures on the Michael Lee Chin Crystal at the Addition to the Royal Ontario Museum, Toronto, ON, Canada, designed by Daniel Libeskind, 2007. The 3D Tekla Structures drawing shows the HVAC as it is worked through the chaotic diagrid geometry of the steel structure.

The Denver Art Museum, CO, USA, designed by Daniel Libeskind, 2006, with M. A. Mortenson Co. contractors, was an early adopter of an integrated BIM system and described in *Engineering News Record*[5] as a "BIM Poster Project". The reflections were: if you can make BIM work on something this complex, then it will be pretty straightforward for more typical projects.

Even to this day, not all 3D models representing a building are necessarily BIM. There are those that contain only visual 3D data, but no object attributes, or those that allow changes to dimensions in one view but do not automatically reflect those changes in other views. These miss the data for supporting the construction, fabrication, and procurement processes. In early days of BIM the software and computing power had limitations on the complexity of the data, so that there are very few building projects that undertook irregular curved geometries at that time.

6. PARAMETRIC MODELING

The issue with the shift from Deconstructivism to parametric design is one of precision. Deconstructivist projects tended to be very angular and some could be even be characterized as rough in form and detailing. Parametric modeling uses the highest levels of precision. While the steel fabrication industry was quite capable of matching Decontructivist intentions and resolution, there are issues in approaching the same with parametrically generated forms. Although 3D-printed steel is being forecast, and could truly come to be able to perfectly match the types of shapes that are being created using parametric methods, we are far from that point in the industry at the present time. As a result, designs that use parametric tools need to acknowledge the real limitations of craft and work within these parameters.

This issue holds for the fabrication of the steel structural system as well as the requirements of the façade. Many traditional façade materials either do not lend themselves to an adherence to curved forms or are prohibitively expensive if produced with precise curvature. This circumstance requires a higher level of design coordination between the structure and the façade. This will be discussed in more detail in several of the curved steel case studies in *Chapter 8*, but the bottom line is that although software can produce true curves, it might be in the interest of the project to fall back to methods of fabrication that approximate curvature through the use of straight and flat segments, approximating irregular curves by seamlessly connecting arc portions of true circles, or by working with scale as a means of negating the need for true curved surfaces. This is discussed in *Chapter 7: Economically Driven Strategies*.

Galaxy Soho in Beijing, China, designed by Zaha Hadid, under construction in 2011. The parametrically driven curvature required adaptation as it transitioned to form the structure and cladding for the project.

The curved skylight of Galaxy Soho was easily handled using a steel lattice structure which permits the use of flat glazing panels while conveying a highly effective impression of true curvature from below.

Modifications were required in the design of the cladding for Galaxy Soho, 2013. Parametric design was used to create the overall curved forms of the building. The cladding panels had to be simplified to work with the capabilities of local industries. The steel structure is concealed, so did not require a high level of fastidious detailing.

7. MATCHING THE SOFTWARE WITH THE EXPECTATIONS OF THE PROJECT

It is critical to ensure that the nature of the geometries of the project match the capabilities of the software used by each member of the team. The more complex or curved the geometries of the project, the more important it is to validate the precision of the software and workflow early on in the project. A high level of communication is critical to the success of the built project.

Most software is able to work with complex angular geometries. Some software still has limitations when it comes to handling curves. Cloud-based systems that offer interoperability between the work of the architect, engineer, steel fabricator and façade consultant are currently preferred.

MANAGING COMPLEXITY

CHAPTER 3

This chapter looks to compare visual complexity to configurational complexity. When a designer seeks to create an aesthetic in expressed steel that embodies complexity, there are many ways and methods to achieve this. These methods will have a direct impact on the overall cost of the project due to associated challenges in fabrication and erection. This chapter sets up the third part of the book that examines projects on the basis of their complex typologies.

1. DEFINING COMPLEXITY

Complexity can be defined as the state or quality of being intricate or complicated. By applying this notion to steel structures we are referring to those systems that deviate from the standard orthogonal structural systems that typified 20th century buildings. A general increase in the complexity of architecture and its structural systems has been the result of this marked shift away from the purity and simplicity of Modernism. Additionally, the perception and constitution of complexity has changed over the same time – from complex in appearance (High Tech) to complex in actual configuration (Deconstructivism) – as has its relationship to structural steel design.

In order to advance our understanding and ability to better design complex steel towards buildability, there are key questions to be asked.

- What exactly constitutes complexity in steel design?
- Are all kinds of complexity the same?
- Are there different ways to achieve or respond to the desire to be complex?
- What does the increase in complexity mean for the design and practice of architecture with specific reference to structural steel design?
- What does the increase in complexity mean for the fabrication and constructability of steel?

We could likely add in here, "and why should architects be concerned about any of this?" As will be demonstrated by the constructed examples in this book, knowledge is empowerment. Better outcomes are possible if a more complete understanding of the interrelationships between complex steel structures and fabrication, detailing and erection issues are incorporated into the overall design detailing and strategies. The base knowledge pertaining to general issues was addressed in *Understanding Steel Design: An Architectural Design Manual*. A detailed exploration of the requirements associated with exposed steel were expanded in *Architecturally Exposed Structural Steel: Specifications, Connections, Details*. There are additional issues that come into play in conjunction with the complexity associated with non-orthogonal geometries that will be addressed in this book. There are different strategies that can be adopted when approaching the detailing of a complex steel structure – differing paths that can be taken towards approximately the same end result that can incur differing costs and amounts of time. Some of these reflect back on the strategic detailing approach addressed in *Architecturally Exposed Structural Steel: Specifications, Connections, Details*, which outlined a matrix of categories and characteristics that reflected ascending levels of fit and finish expectations in AESS as a function of distance to view and the programs of the architecture.

Federation Square in Melbourne, Australia, designed by LAB Architecture and Bates Smart, 2002, uses a chaotic three-dimensional configuration comprised of galvanized square steel tubes to define the expression of the architecture in this important civic project.

2. COMPLEX TYPOLOGIES

It is easier to design for complexity if we recognize its manifestations as having varied typological bases. Not all complex, non-orthogonal structures are the same, though they do share some common technical challenges in their fabrication and erection. As a function of the aesthetic motivations for the project, some will be more expensive or difficult to fabricate and erect. Others might be able to be modified by adopting a different approach to the geometry or fabrication methods, with approximately the same end result. The following represent seven rough typological classifications. Representative projects will be looked at in more detail in Chapters 7 through 11 of this book.

The Sydney Convention and Exhibition Centre, Australia, designed by Philip Cox Richardson Taylor Partners, 1988 (demolished 2013), combines round tubes with a tension system to create a visually complex arrangement of steel that retains a low level of structural complexity.

Mixture of Member Types

Complexity can speak to a shift away from standard concealed detailing and the associated use of hot rolled sections and towards a mixture of member types that would include hollow structural steel sections (HSS, RHS, SHS, EHS), tension material such as rods as cables, and the associated fittings. This would be typified by many High Tech or simple contemporary AESS projects. This type of complexity can also arise from the associated layering of simpler structural systems. This might speak to visual complexity and not necessarily complexity in the calculation or fabrication of the systems. Simple angles and offsets of overlaid systems will cause a deviation from more standard orthogonal systems that may include the introduction of redundancies. The base geometries may still in themselves be quite regular, with the appearance of irregularity arising from the overlay of the systems. Layering is also applied to more complex geometries to intentionally increase visual complexity.

The Seattle Central Library, WA, USA, designed by OMA/LMN, 2004, uses a layered diamond grid to define the shape of the building. The varying geometries and resultant eccentric loading created challenges in the design, fabrication and erection. Large sections of the diamond grid were shop-fabricated to reduce on-site connections, which is a usual approach to the erection of complex three-dimensional structural types.

Three-dimensional Structural Types

From there we can move to include three-dimensional structural types that cannot be solved as simple planar structures. This increase in three-dimensionality can also vary in its degree of irregularity, adding challenges to the problem at all phases of calculation, fabrication and erection. Such types may include box or triangular trusses of varying cross-section. Diagrid-, spaceframe- and lattice-type structures would also fall into this category.

The Lou Ruvo Brain Institute in Las Vegas, NV, USA, designed by Frank Gehry, 2009, uses an exposed steel system to support the undulating façade screen. There is no apparent ordering system. A wide range of member types and connections can be seen.

Chaotic

Some structures are absolutely chaotic and have no recognizable basis in regular geometries or repetitions. Many Deconstructivist projects would fall into this category. Connection design must accommodate numerous odd angles of intersection of varying numbers of components. There is often a great variety in the selection of structural member types, adding complexity to the approaches to connection detailing. The instance of unique geometries is extremely high, negating the economic practices available to modular projects.

The addition of a curved bottom chord on this three-dimensional truss at the Francisco Sá Carneiro Airport in Porto, Portugal, designed by ICQ, 2006, increases the design and fabrication challenges for this element.

Simple Curvature

We will see the introduction of simple curvature as increasing the difficulties in the engineering, fabrication and erection of non-orthogonal structural systems, creating unique problems. The creation of curved elements often depends on very traditional methods of steel-bending that require a higher level of craft in the fabrication and erection processes. These processes are addressed in detail in *Understanding Steel Design: An Architectural Design Manual, Chapter 8: Curved Steel*.

The varying curvature in the roof of the Southern Cross Station in Melbourne, Australia, designed by Grimshaw Architects, 2007, presents a higher level of fabrication and erection challenges than a uniformly shaped vault. Extensive fully remediated welding is needed to effectively recombine the varying curved sections into a credible, consistent whole.

Complex Curvature

Complex curvature is typical in contemporary parametrically driven schemes where the desire is to remove repetition and introduce new parametrically derived patterns. These projects are often the most challenging to build as even the contract documents must shift from conventional 2D drawings and include 3D models to properly convey the relationships and dimensions of the project. Innovative fabrication strategies are required to adapt methods better suited to creating regular curves. More attention must be paid during fabrication to track the individual elements given their unique curvatures.

A large cast node sits at the intersection of eight large concrete filled steel tubes at the Queen Richmond Centre in Toronto, ON, Canada, designed by Sweeny&Co Architects, 2015. These concrete filled steel frames support an 11-storey concrete office tower above.

Cast Connections

Castings that were introduced in the past to add a level of complexity in terms of ornamentation have given way to cast connections whose purpose it is to simplify structural connections. Although designed to simplify load transfer, the inclusion of cast connections often introduces a level of complexity to the project in terms of specialized engineering expertise as well as requiring a specialty caster. Castings are often associated with nodal conditions designed to solve the smooth transfer of forces where a larger than normal number of members connect at a point that tends to be non-orthogonal in nature. Casting methods were addressed in detail in *Understanding Steel Design: An Architectural Design Manual, Chapter 10: Castings*. Nodal connections are presented in more detail in Chapter 10 of this book.

The Peace Bridge in Calgary, AB, Canada, designed by Santiago Calatrava, 2012, is formed using custom brake-formed welded box sections. Fabrication was eased by the predominantly symmetrical nature of the design.

High Level of Custom Fabrication

While most of the previous categories made predominant use of standard steel members, the shift to custom fabrication from plate steel, when combined with irregular geometries, results in the highest challenges for design, fabrication and erection. This type of fabrication can have the greatest variation in costs given that many of the elements constructed have little precedent in terms of fabrication and tend to be quite project-specific. This is particularly the case when working with brake formed steel to create unusual shapes. These projects tend to require extensive use of welding and remediation techniques.

3. THE COST OF COMPLEXITY

Complexity comes at a significant cost premium over standard orthogonal steel construction. Yet this cost can vary greatly as a function of the approach towards the creation of the complex geometries of the structure. Not all projects can afford to pursue higher cost methods. The following are some major factors that influence cost, accompanied by suggestions towards mitigating the same.

1. The base starting point would be the overall choice between angular and curved geometries. Angular geometries can be easier to design, fabricate and erect than more complicated curved geometries.
Alternate paths:
Be prepared to meet with a steel fabricator to discuss your options prior to committing the design fully to a client. If the form is somewhat simplified there can be significant cost savings.

2. The next aspect would be the desire to use curved versus straight members to create a curved form. Bending steel or cutting curved shapes is significantly more expensive than using straight members.
Alternate paths:
You can create overall curved geometries without using curved members. You can make the members straight and the cladding curved or the members curved and the cladding/glazing flat as a function of the scale of the building.

3. This is multiplied by the degree of irregularity in the fabrication of the members and

4. the degree of irregularity of their connections. The more unique situations there are, the higher the cost. It is true that some computer-driven fabrication methods can reduce the cost of unique members over standard fabrication methods. However, the bottom line is that it will cost more in fabrication as well as during the erection process.
Alternate paths:
Can some of the irregularity and frequency of unique connections be reduced? Can the base structure be made more regular and overlaid with a second set of members to give the impression of complexity?

5. How large is the structure and

6. how far is the fabrication shop from the site?
Alternate paths:
Maximizing the amount of shop fabrication can reduce costs over significant welding and finishing operations happening on site. Detail the project to use discreet or hidden splice connections to reduce the requirement for excessive on-site welding. Is the site large enough to set up a temporary enclosed fabrication shop to allow the fabrication of elements that would exceed normal transportation limits?

7. Is the steel exposed? Architecturally Exposed Structural Steel (AESS) has a substantial increase in cost over fully concealed steel. If so, what level of AESS? *(Refer to Architecturally Exposed Structural Steel: Specifications, Connections, Details for more information).*
Alternate paths:
The distance to view of the steel can make a high category of AESS unnecessary. The choices of members and fabrication details can be softened if the material is situated beyond 6m/20ft from view.

8. Is the exposed steel on the interior or exterior of the building – so, is weathering an issue? This will impact the method to corrosion-protect the steel which in turn may suggest different methods of handling the connections and detailing.
Alternate paths:
Exterior steel must be protected from corrosion. If you are using hot dip galvanizing, the members must be sized to fit in the zinc bath. This may require adjustments to the locations of splice connections for large members. If a galvanized appearance is not desired, then a final coating will need to be budgeted or a more expensive epoxy intumescent type used. Stainless steel performs well in exposed environments, but there are limitations on the types of members and connections that are reasonable to expect.

9. Is the steel partially exposed? In many cases the AESS type steel is set behind screened elements. The steel is either partially in view or simply exists as a silhouette.
Alternate paths:
If the details of the fabrication are not clearly visible, the choices of members and connections can be simplified.

10. Do the connections use standard fabrication methods or cast nodes? Both can work structurally, but the cost of castings is normally higher than standard fabricated steel. However, the choice to use castings can be aesthetically as well as structurally driven so may be well justified.
Alternate paths:
If large nodal connections are critical to the look and performance of the structure, it is reasonable to ask the fabricators bidding on the project to provide a 3D prototype to suggest different approaches with varying costs that may not have arisen during the design process.

11. Does the structure use standard structural shapes or are the members formed from custom plate steel? This is significant when looking at the various categories of AESS, as the highest categories make predominant use of custom-fabricated members at a significant cost premium – that is driven by aesthetic requirements.
Alternate paths:
There are various ways to create members if a complete custom approach is unaffordable. Box sections can be made from hollow structural steel instead.
There are various ways to create sharp-looking corners on custom box sections.

If we can agree that complexity is normally an aesthetically driven design intent, then we can open up the possibility of achieving this by a variety of means. I have identified a number of different approaches to creating visual complexity, and the costs vary greatly as a function of the fabrications processes. It is then possible to strategize a series of alternate paths to the creation of the project. These options will be addressed in detail in Chapters 7 through 11.

The BMW Welt Museum in Munich, Germany, by Coop Himmelb(l)au, 2007. A view up into the double cone structure comprised of twisting, all-welded hollow structural AESS. Larger assemblages were prefabricated and then welded together on site, with the fully remediated welds positioned mid-span between the connection points. The lighting in the space assists in concealing direct views to potential splice locations. Advances in software were essential to the development of this project.
For more information on this project see p. 122.

ED-4S
IB-4S
1-800-UR RENTS
United Rentals

FABRICATING THE STEEL

CHAPTER 4

On complex steel projects it is essential that the steel fabricators be included in the project development, as they may have critical suggestions regarding innovative approaches to designing for best fabrication and erection. This is particularly true for connection design, especially the splice connections completed on site, as these must reflect logistical restrictions of which the architect may be unaware.

1. COMMUNICATION AND THE TEAM

Different jurisdictions have varying approaches to handling team logistics and communication. Regardless of the geographic location, complex steel projects require a higher than normal level of understanding among the players. On traditional concealed steel projects it is normal for the architect to be distanced from the selection of the steel fabricator and erector. These decisions will normally be made during the bid or tender process after the drawing set is complete. On complex steel projects, whether concealed or using AESS, the involvement of the fabricator for advice early on during the design phase can facilitate a more successful project. This does not necessarily infer a pre-selection of the fabricator. Fabricators can be retained to assist in this process and later provide bids alongside their competitors.

In some jurisdictions and for some clients it is normal to divide a larger steel contract into subcontracts, with several separate entities to handle the fabrication and erection. In other places, several fabricators might be awarded portions of the overall contract because of the sheer tonnage of steel required. On very large projects, the production capacity is often beyond the plant capacities of a single fabricator. As tight tolerances are absolutely essential when working with complex angular geometries and curves, great care must be taken when subdividing the work. A fabricator may desire to pre-check the fit of the pieces in the shop, and this is more easily accomplished if they have either fabricated or have easy access to all of the adjacent pieces and have them in the shop for test assembling.

Verifying the working method and communication capabilities of the team in terms of software will be critical for coordination. This is a fluid aspect, as software is constantly improved. The requirement of interoperability extends also to associated consultants such as mechanical and electrical, as clash detection is essential when dealing with geometries that are difficult to comprehend and require 3D visualization. Accurate dimensional coordination with cladding suppliers is also essential, as the complexity of forms is translated to the envelope design.

The diagrid baskets for the Brookfield Place Pavilion in New York City, NY, USA, designed by Pelli Clarke Pelli Architects, 2013, were fabricated in Hamilton, ON, Canada and transported hundreds of kilometres to the site. A strategy for breaking the five-storey baskets into transportable elements was critical to the success of the project and greatly influenced site connection strategies.

On complex steel projects, the track record of the members of the team is also crucial for successful cooperation. The steel fabricators and erectors should have experience at the required level of complexity. This is not the occasion to go for the low bidder, and the client needs to understand that sometimes costs saved up front can mean greater expenses down the road to rectify deficiencies.

The delicate nature of the elements that form part of the diagrid baskets for the Brookfield Place Pavilion meant underloading the transport in order to prevent damage to the AESS pieces. In addition, they were shipped in the correct orientation to prevent overhandling and potential distortion during installation. Additional temporary support framing was custom-fabricated to assist with stability during transportation. The connection points of the temporary support systems to the structure will require remediation.

2. MANAGING ELEMENT DESIGN SIZES

It is obvious that the overall structure must be broken into transportable and erectable elements. The most appropriate locations in the structure to make on-site connections, as a function of engineering strength, must be reconciled with very practical connection design issues, which are of concern to the designer as they will directly impact aesthetics and cost. It is helpful to be planning around the maximum transportable element size, as this parameter will impact the connection and splice design decisions, which in turn feed into aesthetic concerns.

Transportation limits are dependent on:

- dimensional capacity of the truck, articulating truck or barge
- weight restrictions for roads and bridges
- bridge and road clearances
- turning widths along the route
- working towards the least expensive transport assistance (following car versus police escort)

Erection limits are dependent on:

- staging area size and location
- site access for offloading and oversized lifts
- crane capacity and type
- crane position
- crane reach

Finish application:

- galvanized elements must fit in the galvanizing bath with a single dip
- elements that only require prime painting and will ultimately receive their final finish on site can be virtually any size
- prepainted elements may require a special temporary facility be built as a function of their size and the nature of the finish

The 27m/88.6ft-long legs for the Addition to the Ontario College of Art and Design in Toronto, ON, Canada, designed by Will Alsop, 2005, were prefinished at the fabrication shop. A special shed was constructed to apply the thick intumescent coating. An articulating truck was needed to be able to access the site, located in a very dense part of the city. The length of the leg exceeded the limit of a standard trailer, and it was essential to complete the entire unit in the shop, as on-site splices were not desired.

The cast node used for a primary connection on the delta frames at the Queen Richmond Centre in Toronto, ON, Canada, by Sweeny&Co Architects, 2015, needed to be test-fit to the large angled tubular legs prior to arriving on site. This required that the node be shipped from its remote point of fabrication to the primary steel fabrication shop. Also, the conical ends of the legs, fabricated at a roller-bender facility, needed to be shipped to the primary fabricator for assembly of the element.

The nature of possible challenges of fabrication in the shop is fairly unlimited. Most fabrication shops have a high bay clearance and adequate lifting apparatus so that it is possible to position, rotate, prop and arrange the elements for ease of access for the ironworkers. Some shops are kitted out with full robotics for handling, cutting and welding.

Even in the best fit-out shop, the type of tasks associated with complex structures can be more physically challenging than during standard construction. Odd angles of intersection can present physical challenges for the welding operations, so that manual welding may be preferred over robotics. Arriving at a clear understanding of fabrication will assist in clarifying its impact on the overall contract costs as well as the required time.

3. CONNECTION DESIGN

Connection design strategies were addressed at length in *Architecturally Exposed Structural Steel: Specifications, Connections, Details.* I will speak in this section to aspects of connection design that feed more directly into the decision-making process for more complex, non-orthogonal projects.

Welding versus Bolting

The choice to weld or bolt a connection has less to do with strength issues, as both methods can generally provide the same strength, and more with aesthetic preferences. Welding provides a smoother look and bolting a more technical one. There are, however, significant differences associated with the methods when looking at logistics, cost and quality. Welding is more adeptly handled in the fabrication shop. Skilled ironworkers and newer robotic systems can provide a highly consistent weld appearance throughout. If welds need to be remediated, ground and filled, this work is far easier to do in a climate-controlled environment where it is also simpler to manipulate the orientation of the element.

Welding Strategies

If a high proportion of welded connections is desired, ensure that the majority of these are situated within the discrete elements so that they can be done in the shop. Limit on-site welding wherever possible if costs are a concern. Think carefully about the style of the connection and its weld. Not all welded connections look the same or require the same level of remediation. Referencing the AESS categories, only AESS 3 and 4 are specified to have all-welded connections and ask for weld remediation. So if designing steel situated at a visual distance of a lower AESS category (beyond 6m/20ft away) no weld-grinding is permitted.

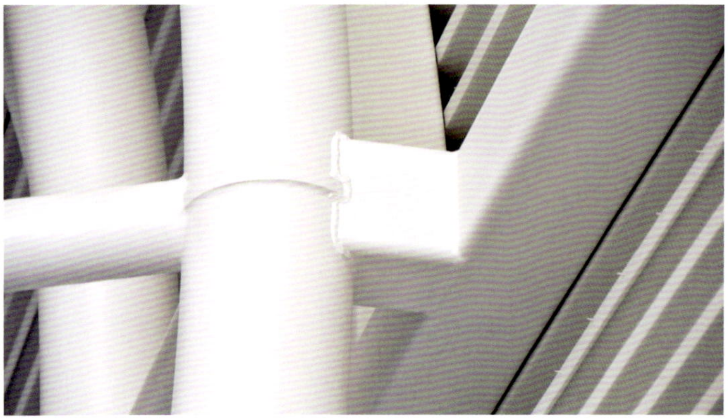

The welding on the left-hand side of the image was done in the shop. The welded connection at the back of the tubular truss was completed on site. As the site connection sits behind the truss, out of view, the decision was rightly taken not to remediate or grind the weld, the location also making these operations difficult to access. Note that the primary chord is larger than the incoming round web tube, allowing for a neat fillet weld with minimal remediation requirements. A plate detail has been used to splice two curved sections of tubular steel, negating a requirement for a seamless weld. This location was carefully coordinated to work with the overall design appearance.

When fabricating truss-like members, setting a hierarchy of members so that tubes taking a higher compressive load are larger than incoming members can allow for the use of fillet welds, provided that there is adequate shoulder room on the primary member. This also creates an easier situation for access to perform the welding. The hierarchical size strategy applies as well to members with circular cross sections and mixed member types so that the fillet weld will not bleed over the edge.

There are some instances where grinding the welds smooth is critical to the proper appearance or aesthetic reading of the connection. This happens regularly when square or rectangular HSS (hollow structural steel) sections intersect in plane – i.e., when a flat X intersection is desired.

The complex connections in the BMW Welt Museum in Munich, Gemany, by Coop Himmelb(l)au, 2007, use two approaches to welding. The rectangular HSS tubes that form the X need to have full weld remediation to provide a continuous appearance on the face of the connection. The tubes that intersect the X are situated slightly behind. This allows for simpler fillet welding that needs no remediation and also results in a layered look, relieving fabrication challenges and creating a more dynamic, hierarchical appearance. This connection was shop-fabricated. The splices between the larger elements occur along the lengths of the tube and have been fully remediated to conceal the connections.

This fabrication image reveals the level of grinding required to create the flat-face tube intersection detail. In order to give strength to the connection, the tubes that are discontinuous through the joint must be bevel-cut to create a suitable void to hold the weld. In this case, the front faces have been held to perfect alignment and the rear faces offset, as they are removed from clear view and can have their detailing softened.

Intersecting round tubes can be handled in varying ways, many of which require little or no remediation if high-quality, consistent welding is provided. The distance to view and the multiplication factor (high number of connections) can justify the softening of the fabrication requirements to satisfy budgets. Round surfaces present a greater challenge to grinding operations. It is quite acceptable to use filler to round out the weld as part of the remediation process to create a smoother overall appearance. As with any repeated connection, it is important to think of the multiplication factor when deciding on an approach. The extra cost and time may be unmanageable if required on hundreds or thousands of connections. The use of robotic welding can be examined where the geometry of the repetition is consistent.

This detail is part of a pedestrian bridge enclosure formed by a diagonally intersecting grid of circular tubular forms. The tubes angled in one direction are continuous, those in the opposing direction cut and welded in place. There are hundreds of points of intersection and so the multiplication factor is invoked. A simple fillet weld is used throughout with no remediation. Although the distance to view is fairly close, the multiplication factor governed the decision.

These round tubes form part of a regular, conical diagonal grid. The regular geometry allowed for consistent straight cuts with visible welds to join. There are hundreds of these connections in the structure so further weld remediation would have greatly increased the costs. Roy Thomson Hall in Toronto, ON, Canada, designed by Arthur Erickson, was completed in 1982, well ahead of the possibility of computer or robotic assistance. The building reflects a high level of pragmatism in the approach to the connection design and execution.

Where circular forms can be precisely cut to tight tolerances from plate material as the operations are computer-controlled, the circular shape of a tube is formed as the material passes through a series of rollers. This results in larger deviations between members, making cutting to precise match a challenge. Plates can be used quite artfully to mediate between the cut round tubes.

Precision-cut plate material was used between three intersecting tubes to create beautiful detailing. The plates protrude sufficiently to allow for fillet welding with minimal remediation. The protrusion accommodates slight inconsistencies in the tube dimensions. Additionally, a shadow line is created that enhances the appearance of the detail.

The interior diagrid on the Capital Gate Tower in Abu Dhabi, UAE, designed by RMJM, 2013, uses a slightly oversized cross of rounded plates to allow for simple angled cuts of the incoming members, with joining via fillet welds. There are extreme variations in the connection geometries of the tower, and this arrangement allowed for uniformity in the plate connector fabrication, as it could accommodate a large range of incoming angles from the diagrid members.

Steel plates have been used to connect the stainless steel tubes on the Helix Bridge in Singapore, designed by COX Architecture and Arup, 2010. The plates have been ground smooth to make them recede from view. The intention was to prevent any crevices in the connections that would promote degradation – in addition to a smooth appearance.

The diameter of the tubular members will impact the choice to use a plate connector as a mediator. Where the tube diameters are smaller, the size deviations proportionately less, and the desired aesthetic more delicate, there is little choice but to use direct member-to-member welding. The expectations for remediation need to be tempered by an understanding that the scale and geometries may preclude access for grinding operations. It is common to use filler prior to painting to smoothen out the transitions.

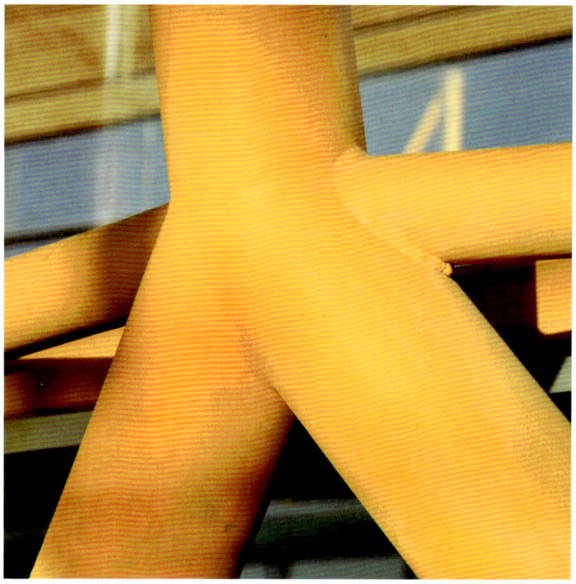

This view of the wind-bracing trusses at Beijing Capital International Airport, China, designed by Foster + Partners, 2008, shows the use of welding with some remediation to manage the diagonal intersection of tubular trusses. A very smooth appearance was desired at the meeting of three members, requiring additional fabrication care. You can faintly see the welding evidence where the vertical members connect to the legs.

An overall view of the wind-bracing trusses at Beijing Capital International Airport shows the delicate scale of the trusses and their member sizes. An all-welded solution was critical to maintaining their light appearance.

Bolting Strategies

Bolted connections are very suited to site work. They are quick to do and allow for efficient use of cranes. Also, there are limited requirements for scaffolds, as access for tightening can largely be managed by the use of lifts that are available on site. Within the basic requirements of AESS, the architect is to specify the side of the connection for the bolt head in order to achieve consistency in the look. There is also an option to ask for a regular Hex Head bolt or a Tension Control bolt, which has a rounded head. TC bolts are not available internationally.

The use of bolted connections can drive the aesthetic desires of a project. Where welded connections are chosen for a smooth and seamless appearance, designs looking for a more technical or edgy aesthetic can choose bolts to enrich the look. Bolted connections are also the logical option if a galvanized corrosion protection system has been selected. When using galvanization, it is advised to select only Hex Head bolts, as they come also with a galvanic coating system. Although Tension Control bolts can be obtained with a galvanized coating, their method of installation results in the removal of a nib from their non-view side, leaving the carbon steel core exposed. These vulnerable pieces must be coated by hand on the site, which might not result in optimal durability or cost.

Splice Connections

The majority of connections used in AESS structures tend to enhance or create design details out of the act of connecting the steel. However, in certain instances, the design intention will be to hide connections or make them less visible. Splice connections are situated internally within elements, as a solution for element sizes too large for transport in one piece. Designers will normally aim to suppress this connection.

The limitations on the transportation size of elements will often require elements to be shipped as a number of smaller pieces, so that site connections have to be carried out. In AESS projects, particularly those designed with an all-welded aesthetic, this can present erection issues, as on-site welding can be problematic and standard bolted site connections are liable to disrupt the visual lines of the elements.

There are three basic approaches that can be adopted to handle splicing steel (typically tubular members) on site, each with varying impacts on cost, time and aesthetics.

Seamless tube-to-tube weld:

- most expensive
- limited to use on AESS 3 and AESS 4 categories (top-end projects)
- requires scaffolding for the ironworker and preheat equipment
- time-consuming as a function of the number of weld passes
- weather-dependent (extra cost to create weather-protective enclosures)

Hidden bolted connections:

- The use of a cover plate over the bolted connection can be made to appear very close to a seamless welded connection.
- less expensive than a welded connection
- quicker to complete than a welded connection
- can be executed from lifts and does not require special scaffolding

Discreet bolted connections:

- least expensive option
- Bolts are visible but the geometry of the connection is held visually "trim."
- quickest to complete so potentially lowest cost
- can be executed from lifts and does not require special scaffolding
- may require specialized equipment to access the bolts for tightening

The Tower of Hope was created using galvanized steel. Large, gently rounded plates were used to connect the wide-flange members. Hex Head bolts were used extensively to create a rougher appearance of the truss members.

Bolted connections were used on the interior of the museum as well for connections between elements, as well as splice connections that were not required to be at all unobtrusive. The roughness of the connection design intentionally feeds into the aesthetic plan. There is, however, consistency in the bolt head placement within the connection points. Even where plates have been used to connect tubular elements, the plates are slightly oversized to permit less expensive fillet welding and add to the overall texture.

The Canadian Museum for Human Rights in Winnipeg, MB, Canada, designed by Antoine Predock, 2014, was looking for an architectural language in the steel that spoke about human suffering. Highly angular geometry combined with extensive use of bolted connections was chosen to contribute to this desire.

It is always important to bear in mind not only the distance to view when deciding on the splice connection type, but also the multiplication factor. If there are numerous splices that may also be somewhat difficult to access, the costs could be prohibitive as a function of the splice method chosen.

Seamless Tube-to-Tube-Welded Connection

The issue when making a splice on a tubular structure lies in the positioning of the splice in the mid-span position. It is unlikely that the designer wishes to highlight the interruption, and so chooses to make the connection "disappear." When this happens on the site, the logistically easiest approach will be to make the connections while the portions of the elements to be connected are sitting stable on the ground, ensuring easier access. The spliced piece can then be lifted as a complete piece. If this is not possible due to a limited staging area, it will be necessary to provide shoring support to the pieces and do the connection in the air. This adds extra cost for the shoring as well as the installation of temporary work platforms. It is standard to add temporary connection plates to the tubes to allow for a quick, temporary bolted connection to hold the elements in proper position. These plates must be cut off when the welding is complete and their attaching points ground and remediated.

Seamless tube-to-tube weld: An ironworker is completing the surface remediation on the splices in a tubular truss. Male and female ends were prepared with temporary bolt plates to hold the elements in proper alignment. Normally at least one tube end will be bevel-cut to allow for the placement of the weld material. The other member has a back stop inserted to limit the weld flow. After multiple passes of welding, the surface will be ground smooth. Filler will be applied and sanded to ensure a perfectly smooth finish. Primer will be applied prior to the application of final finishes.

Hidden Connections

Hidden connections are a very reasonable alternate solution to the seamless welded connection when the elements are at a distant view. This type of connection allows for a quick but permanent bolted connection to take place at the time of the lift. Access must be provided to install the cover plates and remediate their joints to the point of near invisibility if desired.

Hidden connection: A fairly standard bolted plate connection is welded to fit completely inside the tube. A pair of curved plate sections are tack-welded over the gap in the tube. The joints can be filled and sanded prior to painting. If this connection is located in an exterior location, care must be taken to completely seal the plates to prevent water from entering and causing hidden corrosion.

Discreet Connections

Discreet connections are an artful way of designing the splice connection so that the bolts are exposed, but the overall architecture of the joint is very trim in order to maintain the visual lines of the joining members. This attitude can be applied to any type of joining members, be they tubular or standard structural shapes. The examples shown illustrate a range of approaches with a variety of shape types. What is consistent is the use of exposed bolts. Care has to be taken when these are located in a weather-exposed location to ensure that water cannot accumulate or puddle in the connection.

Splice Connections were illustrated comprehensively in *Architecturally Exposed Structural Steel: Specifications, Connections, Details, Chapter 8: Specialized Connections.*

This connection is located on the exterior bracing system of the Leadenhall Building in London, UK, designed by Rogers Stirk Harbour and Partners, 2014. The interior of the square tube is capped to force water to drain from the void that is left to facilitate the installation of the bolts, as well as to partially hide them.

These connections allow for the splicing of large arched elements at a transit station in Sydney, Australia. As standard bolt-tightening equipment will not fit, a special implement is used and the pockets are sized to permit access.

The structure on Federation Square in Melbourne, Australia, designed by LAB Architecture with Batesmart, 2002, is comprised of a chaotic assembly of square tubular elements. The slim-slotted connections that have been fabricated from plate steel allow the bolts to sit on the faces of the tubes. The technical look is quite compatible with the choice of a galvanized finish for this building. The structure is partially exposed to the elements. The welded connections within the elements have also not been remediated, as their texture is desired as an accessory to the overall aesthetic.

The steel diamond grid frame for the Seattle Central Library in Seattle, WA, USA, designed by OMA/LMN, 2004, is comprised of wide-flange (Universal) sections. The elements have been fabricated to the maximum transportable size in the shop. The site connections are all bolted, fitting neatly inside the members and using a typical tension splice connection. The numbers of these splices and the angular orientation of the grids would have made a fully welded connection prohibitively expensive due to shoring and access requirements to complete the work. Connecting plates have been welded to the flanges to cover the gap and provide for visual continuity.

4. THE FUTURE OF FABRICATION

This book has been written with an emphasis on practical techniques that are extremely applicable to current methods and upcoming abilities of most fabricators. Although 3D-printed footbridges and smaller steel elements have been produced, the current costs and technical limitations do not present a viable alternative to fairly conventional fabrication and erection methods.

Robotics, on the other hand, are realizing great improvements in the fabrication processes of some fabricators. Much like the evolution of the architectural office at the advent of CAD, it will take some time for fabricators to first see the need for this tool, and secondly, be able to afford it. Current robotics enable a hands-free shop, where the entire process from offloading of the truck to finished welding of elements can be done via simple barcode tracking. Steel can be handled with precision and enough care – with some modification of the steel mesh belts used to turn the steel – to allow even the handling processes to feed into AESS projects. There is an ability to more quickly and consistently process identical pieces, as well as to accommodate variations. These capabilities should allow for some economies in complex steel projects, provided the details work within the limitations of the welding angles and processes within the systems. When designing complex angular steel projects it would be prudent for the designer to understand the capabilities of the fabricators bidding on projects, including a tour of facilities to understand strengths when deciding to award work.

Even with the boost provided by robotics, the fabrication of many curved elements lags behind. With the exception of computer-controlled cutting of plate, any bending processes remain highly manual in their implementation and precision checking. As curved steel applications are outside of the norm for most projects, there is at present little impetus to improve on these systems. 3D printing will offer great benefits to such shapes in the future, but it is not a viable alternative at the present time.

CORROSION PROTECTION & FINISHES

5 CHAPTER

To act responsibly in the design of steel structures exposed to weathering means understanding the limitations of coating systems and altering the design accordingly to ensure minimal environmental degradation and reasonable maintenance requirements.

1. AESS AND ITS IMPACT ON FINISH SELECTION, CORROSION PROTECTION AND MAINTENANCE

Although complex geometries necessitate much tighter tolerances than standard structural steel that is concealed in order to ensure the proper fit between members for structural integrity, complex Architecturally Exposed Structural Steel will have even tighter tolerances than complex steel that is to be concealed either entirely or partially. The accuracy of fit between adjoining members and the ability of the fabricated connection to identically match the digitally produced drawings is critical to ensure not only a good-looking final product but also to not slow down the erection process due to fitting issues.

AESS also requires much more careful handling than standard structural steel. This care in handling begins in the fabrication shop and the preparation for shipping. In some cases, the elements may come prefinished, ranging from a simple primer coat, to galvanizing to a final intumescent coating. Even if shipped with only a protective primer coat, care needs to be taken not to damage the surfaces, as any abrasions will require remediation on site. Large elements will require additional temporary support in transport in order to prevent deformation. While the designer is not involved in the design of the implements used to assist shipping, it is important to understand that this handling will impact the overall cost of the project. More delicate elements will need to be shipped separately, meaning increased transport costs over standard steel.

The Helix Bridge in Singapore, designed by Cox Architecture and Arup, 2010, is fully exposed to a harsh marine climate. For this reason it was decided to fabricate the bridge from Duplex 2205 stainless steel, the highest grade of stainless material, as lower grades would not have been as durable. This choice of material required extensive welding and weld remediation to complete the aesthetic.

When designing with AESS it will be important to discuss the nature of the final finish and where this is applied (shop versus site). In some cases this will impact the size of the elements and impose splice connections on the design. It will be important for the ultimate durability of the project to ensure that coatings are applied in optimal conditions.

AESS should always start out with a professional blast-cleaning to remove any mill scale, grease, et cetera, that will prevent the proper adherence of the finish. This is a cost above the cleaning required for standard concealed steel.

If not galvanized for a permanent exposure to weathering, AESS is normally primed to prevent excessive oxidation on site and preserve the integrity of the care given during shop fabrication. The primer will be held back from areas that are adjacent to future welded connections. These will be cleaned up immediately before the weld operations take place. The primer must be coordinated with the final finish to ensure compatibility.

2. PROTECTION METHODS

There is a wide variety of methods to protect the steel from environmental degradation and fire. These choices are most critical to the design and detailing of the structure when it constitutes Architecturally Exposed Structural Steel, and even more so when detailing highly complex angular or curved structures. There have been substantial advances in protection systems since the High Tech period. Many projects constructed during the 1970s and 80s used simple painted finishes that have not proven to be durable, resulting in the degradation of a number of high-profile projects. This is quite unfortunate, as the repair costs associated with refinishing to a more durable state are often prohibitively high, leading property owners to contemplate demolition as the most economical choice. Repair and refinishing is exacerbated when the steel detailing is very fine, as aggressive methods of paint removal, such as sandblasting, cannot be used because they are liable to remove the steel and undermine its strength. Thinner steel products tend to lose significant strength via oxidation.

For exterior steel the main choices would be:

- carbon steel that has been galvanized and left exposed (the galvanized coating is the finish)
- carbon steel that has been galvanized and given a painted finish (the galvanized coating must be properly aged prior to that application or the paint will not adhere properly)
- an epoxy intumescent coating used over carbon steel (also works as fire protection)
- stainless steel
- weathering steel

Stainless steel would be the most expensive and most durable option but comes at the highest cost. Stainless material has a somewhat limited range of sizes and section shapes, as well as requiring specialized engineering expertise, as its behavior is different from carbon steel. It also must be fabricated in dedicated shops, as contact with equipment that has been used on carbon steel can embed small particles into the stainless steel, causing rust spots. There are different grades of stainless steel, as different climates present varying degrees of harshness. Marine climates tend to be the harshest and will require higher grades. The Helix Bridge in Singapore (see p. 44) is an example of a highly complex structure where the choice to use stainless steel will ensure a high level of durability and low maintenance.

Weathering steel is also very durable but not found in many complex steel projects because of the extremely limited choice of material sizes and shapes. You will find it used in AESS 4-type projects, which are typically comprised of custom steel shapes using welded plate material.

The ongoing maintenance of the Centre Pompidou in Paris, France, designed by Piano and Rogers, 1977, is a significant burden. Simple painted finishes are not durable. Urban environments also stain white colors very quickly.

Waterloo International Terminal, London, UK, by Nicholas Grimshaw, 1993. The legacy of extremely finely detailed, fragile, complex structures from the High Tech era presents us with a case for thinking closely about the inclusion of durable finish systems for complex exposed steel projects. Many of these in England are under threat of demolition and seeking protection under a historic status. Many of these lightweight structures were not designed with longevity and maintenance in mind. We have significant information about what does not work well, indicating that simpler painted systems should be avoided. The station was closed in 2007 and reopened in 2018 after a large restoration and renovation.

The Webb Bridge in Melbourne, Australia, designed by Denton Corker Marshall, 2005, is a highly complex structure comprised of lightweight curved steel elements and situated in a corrosive marine environment. The steel used in the design was deemed too light and articulated for hot-dip galvanizing. It was instead given a Polysiloxane coating. Sadly, significant degradation resulted in less than a decade. The finish is weather-aging and peeling, and the fastening system is corroding. The galvanic coating on fasteners is easily damaged during their installation, creating a weak spot in the system. Given the sheer complexity of the design and its location over water, maintenance is difficult and refinishing is prohibitively expensive.

The articulated steel armatures that support the glue-laminated timber façade of the Addition to the Art Gallery of Ontario in Toronto, ON, Canada, designed by Frank Gehry, 2008, use a combination of hot-dip galvanizing and a heavy-duty zinc paint coating. The steel support arms for the curved glulam members were not only complex in design and thereby in danger of heat-induced deformation but also exceeded the capacity of the galvanizing bath. The coloration of the two systems is fairly close, and their location slightly behind the dominant façade made the differentiation in finish colors manageable.

Galvanizing

Galvanizing is chosen for use on exterior steel whose overall aesthetic is technically driven. The structures must be detailed to avoid puddling of rain and snow, allow for good drainage and ensure that all attaching mechanisms such as bolts are also galvanized and their surfaces not unduly damaged during the installation process. Hex Head bolts are preferred over Tension Control (TC) bolts, as the installation process of the TC bolt removes a nib at the back of the bolt, exposing a diameter of carbon steel. If these are not properly sealed with a high-quality zinc-based paint, oxidation will occur. It is also important to note that the appearance of the steel will be somewhat inconsistent. The mottled appearance varies as a function of the thickness of the steel as well as its chemical composition. This will infer putting an intentional "break" in the design at points where steel from different batches is joined. This will be very important to consider on splice connections, whose continuity is critical to the appearance of unity of the element.

Galvanizing will always take place in the galvanizing shop, usually of finished elements that need to be sized to fit in the zinc bath (double dipping end-to-end is not to be done). Hollow structural steel elements will be coated inside and outside in order to prevent accidental interior oxidation from trapped moisture. This will add to the cost of the contract, as the surface area to be coated is approximately double. The element design needs to be coordinated with the galvanizer, particularly in the case of hollow sections, as it is critical in the dipping process that no gasses are trapped during the dipping process or the element may explode. Fabrication will leave small holes in strategic places to allow the zinc to flow through and the gasses to escape.

Keep in mind that when using hot-dip galvanizing that the final appearance will vary as a function of the steel. Detailing feature connections between elements will work to downplay the inconsistencies. This requires that the connection is located between elements. A discreet bolted connection has been used to splice the column, as the complete member would not fit in the bath. The slight variation in hue of the finish is downplayed by the connection detail.

The site-applied zinc-based coating on the Shenzhen Bay Sports Center in Shenzhen, China, is suffering massive degradation. This photo was taken only five years after the project opened. The custom box sections of steel arrived to the site primed, with only their weld-ready edges exposed for on-site welding. The restoration of this system will be a massive undertaking given the scale of the building.

The galvanizing bath is extremely hot (about 450°C), so this can impact the design of complex members, as thinner steel elements may deform with the heat, thereby negatively impacting their potential fit as well as aesthetic quality. It might be necessary to increase the wall thicknesses of some members to compensate. Thicker steel also bonds better with the hot zinc, thereby providing a superior sacrificial coating. Steel is priced by weight, so this procedure impacts the cost of the contract, but it might still be the most prudent course of action for a quality product.

In some instances the detailed nature of the steel is not suitable for galvanizing (too large, too complex, too thin) and will require an alternate solution to provide corrosion protection. Various zinc coating systems exist, but some of these should be approached with caution, as they may either be significantly more expensive than galvanizing (as in the case of metallization) or simply inferior (as in the case of some zinc-rich paints). It is advisable, prior to agreeing to an alternate protection method, to ensure the qualification of the product and applicator, as well as to inspect other projects and their durability over time with the same coating systems. It might even be preferable to rethink the element sizes and splice details to allow for galvanizing or even the use of stainless steel rather than use a coating system that has limited proven durability.

Intumescent Coatings

Intumescent coating systems come as a variety of applications. Although the intention of the coating is to provide a level of fire protection, some epoxy-based systems in addition provide excellent weather protection for exterior applications. Many AESS projects will choose an intumescent coating system as part of the fire protection system.

There are several aspects that need to be addressed when specifying the type of intumescent coating, which in turn feeds into member and connection selection and design strategies. Intumescent coatings, while giving the impression of a painted finish, are thicker than paint and have a textured finish. The required thickness varies with the level of fire protection as well as the thickness of the steel. The thinner the steel, the thicker the coating is required to provide equal protection. The cost of the intumescent material is fairly proportional to the thickness and the number of application layers. In some instances it may be more economical to use thicker steel to keep coating costs down. This will impact the member choice as well as the type of connection detailing. There is little point in pursuing finely detailed, intricate connections, as they will be less defined when a thick coating is applied. It is advisable to consult, early in the design process, with the fire engineer to determine the approach to fire protection.

More information on Coating Systems can be found in *Architecturally Exposed Structural Steel: Specifications, Connections, Details, Chapter 5: Coatings & Protection.*

Site Access to Apply Finishes

If final finishes of any type are to be applied on site, there needs to be a strategy to address the process of application of the finishes together with the environmental protection requirements. While highly creative structural forms may be easily generated during the design phase, and even be able to be erected by crane and accessed by ironworkers (with difficulty) to complete welding and bolting operations, painting and finish applications require access to the entire structure, not just the points of site connection. This is incredibly important to consider for those portions of a structure that are exposed and at great heights.

The final intumescent coating is being applied using a roller on the diagrid baskets at the Brookfield Place Pavilion in New York City, NY, USA, designed by Pelli Clarke Pelli, 2013. The worker must be able to reach and evenly apply the materials. This was done using scaffolding and lift trucks. The material arrived to the site primed.

The bottom of the 27m/88.6ft-long steel legs for the Addition to the Ontario College of Art and Design in Toronto, ON, Canada, designed by Will Alsop, 2005, have been refinished. There is wear and tear on the thick intumescent coating and the cost to re-coat the entire leg is prohibitive. It is a good idea to keep this in mind when detailing large components. Color matching is difficult in time due to the availability of materials and UV fade impact. It is better to detail a reveal or surface change in order to make the color difference less noticeable. Intumescent coatings require a hard-top coat for added durability and control of fingerprints and marking.

Some issues will be different for interior and exterior applications. Interior applications will normally benefit from climate control during painting operations. If the project is looking for a sustainable certification there are likely to be restrictions of VOC emissions. This will impact the nature of the painted or intumescent coating. Water-based intumescent coatings are more sustainable; however, they require a less humid environment and take longer to cure.

Exterior AESS will often arrive to the site prefinished, as these structures suffer with greater exposure to the elements during the construction process. Access on site to apply coatings will be particularly difficult for oversized members. Additionally, as a function of the time of completion, elements may not arrive in the most suitable weather. It is easier to touch up connection points, nicks and scratches from the erection process than it is to coat the entire structure on site.

The highly energetic exposed steel structure at the TGV Station at the Charles de Gaulle Airport in Paris, France, designed by Paul Andreu, Jean-Marie Duthilleul, with engineering by RFR and Peter Rice, 1993, presents extreme difficulties for access for cleaning and maintenance. The high portions of the bow trusses cannot be reached by equipment on site and so continue to accumulate dirt as a result of the open-air nature of the station in combination with the particulates generated by the operation of the trains. The top of the steel is viewable from above.

3. CLEANING AND MAINTENANCE

It would also be prudent at this point to contemplate strategies for lifelong cleaning, maintenance and refinishing the steel. Initial touch-ups during construction have the benefit of access by lifts and cranes that are already on site. If the original application of finishes on site requires special equipment such as lift trucks and scaffolding, then thought needs to be given to this inevitable requirement for long-term care. There are numerous exposed steel projects that have never been cleaned or repainted in many decades due to issues with access.

When considering the color and texture of the coating system, it might be helpful to take into account the nature of the environment. Some urban environments are more polluted than others, causing early staining and degradation of the finish. Train and subway stations and parking garages generate black particulates as a result of their operations. White tends to highlight soil more quickly than muted tones. Simply choosing an alternate color can save cost down the road in terms of the frequency of cleaning requirements. Simple intumescent coatings will require the application of a more durable top coat as their natural finish tends to mark fairly easily. Highly textured finishes trap particulates and preclude standard simple cleaning methods. If the steel is viewable from all angles, consider smoother, easier-to-clean finishes.

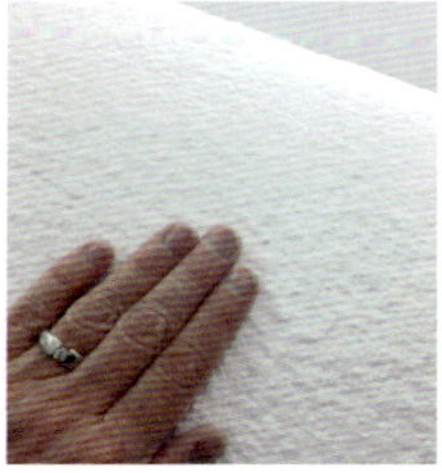

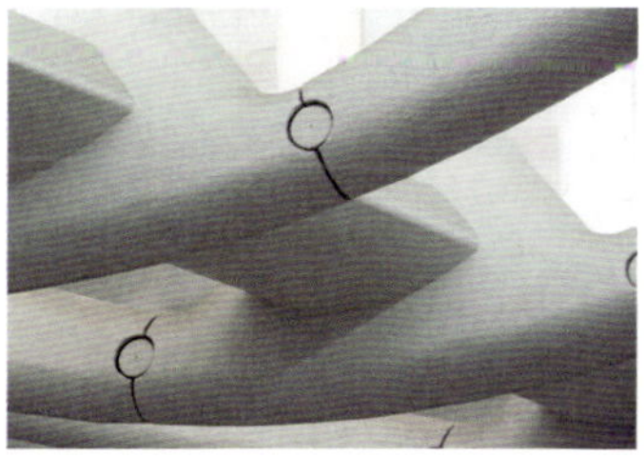

The highly textured white finish that is a critical part of the fire protection system for the Oculus and Path structures in New York City, NY, USA, designed by Santiago Calatrava, 2016, are already showing signs of degradation. The textured surfaces encourage the accumulation of particulates and are difficult to clean. The structure is viewable from multiple angles, providing no means to conceal the grime.

A custom cleaning device has been integrated into the large custom structural steel arches that support the glazing system for the Greenhouses at Marina Bay in Singapore, designed by Grant Associates with WilkinsonEyre, 2012. The platform allows for a reach to the glazing and rotates to maintain level as it travels over the building.

Access for cleaning and refinishing needs to be discussed with the client, as it impacts operational costs. Publicly funded projects tend to have more restrictive budgets and therefore, in spite of an energetic initial investment in a highly expressive exposed steel structure, might not have adequate funds for maintenance.

In some instances the steel may be accessed via the same methods used to provide ongoing cleaning for the glazing. Many innovative systems have been developed to provide ongoing cleaning to structures with complex, non-orthogonal forms. Again, this works best when integrated into the design concept for the structure.

ERECTION LOGISTICS

CHAPTER 6

There is little that is straightforward when it comes to the erection of complex steel structures. Be the elements angular or curved, there will be a constant fight with gravity due to eccentric loads. The team working on the site needs to be highly skilled, as the problem-solving skills and experience required to ensure success are extremely high. It is imperative that designers appreciate the difficulties that must be overcome when erecting these heroic structures.

1. FIGHTING GRAVITY

Although simple in concept and symmetrical in geometry, the delta frames on the Queen Richmond Centre in Toronto, ON, Canada, designed by Sweeny&Co Architects, 2015, was not without its challenges during erection. Fit tolerances between the central cast node and the sloped tubular steel legs were very tight, as a seamless welded connection between the members was the aesthetic goal. The casting design was executed by CastConnex and the steel by Walters Group, a division of labor that required close coordination to ensure proper fit.

There is little about complex steel projects, due to their non-orthogonal geometries, that does not impact every stage of the design, fabrication and erection processes. However, when it comes to the erection processes, many of the variations from the installation of conventional steel present extreme challenges to the project schedule and the skills of the ironworkers on site. Conventional steel projects are largely comprised of symmetrical or balanced elements. In the case of columns, the crane lowers the piece into position and the largest challenge is the alignment of the plate holes to the receiving bolts. Eccentric forms will have odd requirements for lifting. Even having the crane attachment points slightly out of position can make for failed attempts and repeat tries. This will obviously increase the time needed to properly install the element. Although in an ideal world, particularly one that is now so engrained with digital computation, one would imagine all of the lifting points would come pre-calculated, this is not always possible in practice. This is where experience comes into play, and it is usually the lead ironworker on the project who is doing the estimation for the lift points.

Where projects exhibit a greater degree of geometric regularity, it is possible to pre-plan the crane attachment points and the geometries of the chain arrangements to facilitate predictable and accurate lifting. This becomes more critical on AESS projects. If the elements do not align it becomes difficult to manipulate them into position without damaging the materials. When lifting and providing shoring or temporary supports for AESS it is necessary to also provide padded slings and pads for any supports that come into contact with the steel to prevent damage to the surfaces.

The staging area to allow the offload of steel and subsequent pre assembly of elements for the construction of the Addition to the Royal Ontario Museum in Toronto, ON, Canada, designed by Daniel Libeskind, 2007, was generous in urban terms. It was essential to the success of this project that boasts not a single vertical column.

The lifting points for these asymmetrical elements on the Royal Ontario Museum were assessed element by element, on site, as the lift progressed. It was critical that the element arrived to its location at precisely the correct angle of incline, as normal methods of encouragement would be insufficient to fight the weight of the steel with gravity opposing. This steel was erected between 2005 and 2006 by Walters Group. Software of the time was not able to prepare lifting point locations for a project with all unique elements.

The sloped steel legs at the Queen Richmond Centre had their lifting arrangement pre-calculated to ensure precision. Attachment points were welded to the top and bottom of the legs and the chain lengths were set to allow for a single crane hook attachment. Ropes were used by the receiving ironworkers to rotate the element into position.

Ironworkers receive the sloped tubular leg. Both the angle and the rotation were critical to position the piece correctly over the sloped bolted base. The structure was to be concrete-filled for strength as well as fire protection. A Tremie tube opening can be seen, which was used for pumping the concrete into the structure.

2. STAGING AREA

Standard orthogonal steel structures typically do not need a large or oversized staging area, as the members and elements that are brought to site typically do not require further assembly prior to lifting into position. The need for a larger staging area is unique to complex steel projects and will have an impact on the nature of the site splicing and sub-assembly that is required or possible. This can be a major issue on dense urban sites, where land is at a premium and where road closures may be necessary to facilitate erection from time to time.

As previously discussed in *Chapter 4: Fabricating the Steel*, larger elements are broken down into smaller transportable elements and will require splicing on site. This can preferably be done on the ground prior to the lift, as it makes the connections much simpler and safer to access than performing the same operation many meters above ground. It does require, however, a sufficiently large staging area to offload the steel and set it up properly for the pre-lift assembly. It will likely require some temporary shoring to hold the steel elements in the correct position to complete the work. This is normally customized for every project as needed.

Understanding the site constraints will factor into decisions regarding the specific nature of the splice connections specified. There may be insufficient area available on dense urban sites to pre-assemble the elements, forcing the splices to take place after the lift has been completed. In this case, the pieces must be held in the correct position to ensure stability and precision until the connecting operations have been completed.

The relationships between the offloading area, staging area, crane location(s) and the project itself become a question of logistics. Project management prefers to reduce the number of cranes on site, as these come at a very high cost. So it will be up to the erectors to examine the site and come up with a plan. It will help if there is a good working relationship between the fabricators and erectors. If the project is very large and the work awarded to multiple fabricators and erectors, it will be prudent to ensure that areas of connection with extremely tight tolerances are well communicated. This will likely involve site measurements to ensure that the as-built configuration precisely matches the digital drawings.

The stiffness of the members and joints of the Royal Ontario Museum was sufficient to allow the construction to proceed with virtually no shoring requirement. A few tension cables were used for stabilization at discrete points, but these offered little interference to the erection process. This project was considerably different from the Denver Art Museum, also by Daniel Libeskind. The slopes on the members were more extreme on the Denver Art Museum, and a large number of supports was needed to hold the steel in place and prevent deformation during construction. The street beside the museum required closure for a significant period of time. This would have not been acceptable for the Royal Ontario Museum in Toronto, given the prominent urban location of the project. It was possible to construct the project with a single tower crane. Smaller cranes were used from time to time to assist.

The close to nonexistent staging area at the service core side of the Leadenhall Building in London, UK, designed by Rogers Stirk Harbour and Partners, 2014, drove the decision to completely prefabricate and prefinish all of the (yellow) steel elements that comprise the external service core. This architectural office, instrumental in the development of High Tech architecture, developed a method of connections that either makes a beautiful detail of the joint or uses discreet bolted connections where the connections are less visible. The location had very limited access for steel delivery for the service core as well as for the primary diagrid structure.

A lack of at-grade staging area, coupled often with a limit to nighttime-only deliveries, can begin to drive more innovative solutions, for example, prefabrication of more modular solutions, where all of the on-site connections are designed to be done quickly and permanently – unlike the use of temporary bolted connections that precede welding. If the elements can be lifted into position directly from the transport, that provides an even more optimal solution.

3. SHORING AND TEMPORARY SUPPORT SYSTEMS

One of the challenges in the erection process for eccentrically loaded steel is the task of keeping it stable until all of the temporary and permanent connections have been completed. As mentioned, one of the key advantages to completing splice connections in the staging area prior to lifting is the avoidance of shoring and support towers in addition to lessening the need to provide a stable working surface for welding at height. The type of shoring required will in part be a function of the height of the steel as well as the loads to be supported.

Shoring and scaffolding are not only an expense but can also obstruct the access of trades and equipment. Some non-orthogonal structures, such as diagrids, are designed with enough stiffness in the members during the construction phase to be self-sufficient and have no need of temporary supports.

Cables are often used to stabilize the steel, particularly when support is needed for inclined and eccentrically loaded elements. It can be less obstructive to the work space than towers or denser shoring. Cables must be clearly marked, as they present a safety hazard as they are not easy to see.

Steel support towers are required to shore the erection of the curved steel and glue-laminated timber arches at the Richmond Speed Skating Oval in Vancouver, BC, Canada. The 103m/338ft arch is comprised of four 26m/85ft segments. The composite members were fabricated at George Third & Son. The maximum size of each section was based on shipping restrictions. The base sections of the arch were erected and supported by steel towers. To cut down on high-level splice requirements, two sections comprising the center of the arch were spliced on the ground prior to lifting. This reduced the number of towers required. Specialized lift straps were fabricated to fit the V-shape of the section, carefully padded to avoid damage to the AESS members. The chain lengths were adjusted to bring the curved member into the adjacent connection points at the correct angle of incline. The staging area was located within the footprint of the arena, conveniently positioning the sub-assembly of the arch elements in the proper position for lifting.

A temporary support system was used at the construction site for the Richmond Speed Skating Oval to allow the splicing of two large curved beam elements in the correct orientation for lifting. Hidden bolted connections facilitated the splice. The connection will sit at a great distance over the ice rink so that it will not be visible.

The use of towers is essential to the installation of long-span structures such as this square box truss spanning across the Rogers Centre Ice Rink in Edmonton, AB, Canada. The center part of the arch was lifted in one piece and is joined using bolted connections. The nature of the building and the distance to view allow for a softening of the detail requirements from an aesthetic perspective.

The diagrid steel structure for the Royal Ontario Museum was designed to require only a cable support system to ensure alignment while the structure was being erected. It resulted in minimal interference to the site operations.

4. LIFTING THE STEEL

Lifting complex steel elements comprised of non-orthogonal, eccentrically loaded or curved shapes presents a high level of challenge for erectors. Lifting is ordinarily carried out via tower cranes or mobile cranes with a capacity determined by the tonnage and arm length. Innovative methods are required when the element is too large, too eccentric or too heavy to lift with a single crane.

Site access needs to be understood when designing structures with challenging lifts. Access to this location required the closure of an adjacent street. This could only be arranged during late evening hours, meaning that the mobile crane could only be brought to the site immediately prior to the lift. Additional lighting was also required.

Cranes, in particular, represent a very high cost to a project. When the erector is laying out the site, they will try to locate the cranes to minimize the number required by maximizing their reach. This must also be coordinated with local authorities in determining the particular street access for delivery and offloading of the steel. This is of great concern on dense urban sites, as permanent lane closures are disruptive to the traffic flow. Often streets may need to be closed to permit significant lifts on a one-off basis.

The majority of steel components will be sized to be lifted by a single crane. As discussed, there may be challenges to the geometry of the chain attachments for imbalanced elements, but generally the weight of the component and the reach of the crane will be manageable using a single tower or mobile crane (see John Street Pedestrian Bridge, p. 60). On the rare occasion, the location or weight of the element may require a coordinated lift by multiple cranes, chains, winches or hydraulic jacks.

Some multiple lift systems can be computer-controlled to ensure precise coordination. This was the case for the Wembley Stadium Arch (see p. 62), which used five winches to rotate the site-assembled arch into position, and Gardens by the Bay (see p. 106), which used multiple hydraulic jacks to lift the tops of the "supertrees." For the Puente de Luz pedestrian bridge (see p. 58), four mobile cranes were required to simultaneously lift the separate bridge sections to permit fast site connections and obviate the need for temporary support towers.

When erecting steel for bridges over streets, rail lines or rivers, it will be necessary to ensure that there is no danger to those below, which usually means shutting down traffic for the time of the lift. When traffic of any sort must be disrupted for many hours at a time, then the erection process is even more inclined to prefer bolted connections, unless the element can be lifted in one piece. If this causes a loss of income, as in the case of disrupted rail lines, compensation to the affected parties will be required.

The pods for the Leslie Dan School of Pharmacy, Toronto, ON, Canada, designed by Foster + Partners, 2006, were fabricated on the ground. The ground location could not be sited directly beneath the final location, so crane rails were installed to permit a multiple chain lift that was manually controlled while the large element was shifted laterally into position. Given the overall construction schedule, crane access overhead was not possible for this lift.

The diagrid steel baskets for the Brookfield Place Pavilion in New York City, NY, USA, designed by Pelli Clarke Pelli, 2013, were lifted into position using mobile cranes. As this was a reconstruction over an existing building to replace an entrance that was destroyed during 9/11, the location of the cranes was severely limited, as they could not sit on the ground-floor slab. This impacted the logistics of delivery, offloading and the lifts, as the crane reach had to be long and overturning properly countered in weight.

The 51.3-tonnes custom steel node used on One Manhattan West in New York City, designed by Skidmore, Owings & Merrill LLP, 2020, required two mobile cranes to facilitate a tandem lift to move the node from a position that was difficult to access on the lower level of the site, into its final position. The site is situated over rail tracks and so the cranes could not be positioned closer for a single crane lift. This lift could not be computer-controlled and required a high level of communication between the ground and the two crane operators.

5. WELDING AND CONNECTION PROCESSES ON SITE

As discussed in *Chapter 4: Fabricating the Steel*, the objective is to maximize the share of connections that are fabricated in the shop and minimize the share of those completed on site. Erectors have a preference for connections bolted on site, as they are quicker to execute, requiring less time from ironworkers and decreased crane time. By contrast, many projects may have a preference for all-welded connections, so that safe access platforms must be provided. Welding requires pre-heating the steel. The thicker the steel, the more time this requires. Additionally, when making substantially loaded connections, multi-pass welding will be needed. This can take hours if not days to complete, depending on the connection configuration. If the steel is finished to a high level of AESS (3 or 4), good access for grinding and filling operations is necessary.

A view up the Brookfield Place Pavilion showing the extensive multi-storey platform that needed to be constructed to provide the ironworkers with easy access to facilitate the hundreds of all-welded connections for the AESS steel baskets. The welding and remediation took close to three months to complete.

All of the connecting plates that provide temporary support during erection will need to be removed and their locations remediated after welding.

Condition of the welded connections at the Brookfield Place Pavilion following weld remediation: the vertical areas of oxidation, which occurred where the temporary bolted connection plates were removed, require extensive grinding. The connections will be further cleaned and primed prior to the application of the intumescent coating system.

Weather protection can also be a requirement during finishing operations to ensure the best product. This is critical in rainy locations or those that experience snow and cold temperatures. Welding is an exacting process and for the best quality output the ironworkers need to have ergonomically appropriate access and adequate warmth for manual dexterity. The enclosures also help to retain the heat during the pre-heat phase.

A platform was erected around the nodal connection at the Queen Richmond Centre. The multiple-pass welding required for this major load transfer point took a significant amount of time.

Temporary bolting plates are attached to the node and tubes to allow for a quick release of the crane after the lift. When these are used, more surface remediation is required. The alternative would have been creating a substantial support system to precisely hold the four legs which would have had to remain in place for many weeks until the frame was self-supporting. With the use of temporary bolts and additional shoring, there was sufficient strength to proceed with the erection of the four upper legs.

A view of the remediated connection at the point where the hollow steel tubular supports join the cast node. It was critical for the clean geometry that the final appearance was smooth. The connection will be further cleaned to remove all oxidation and primed prior to the application of the intumescent coating. The tubes are concrete-filled to serve structural and fire protection requirements.

6. SAFETY ISSUES

Steel erection is one of the most dangerous trades on a construction site. The safety of the workers must be a top concern of the entire design team. Digital drawing tools enable architects to create wondrous, complex designs, but not all proposals feed into a safe work environment. Complex steel structures, chaotic geometries, large curves, high spans and oddly shaped buildings create very challenging scenarios when it comes to access for the ironworkers to complete connections on the site.

There needs to be a thoughtful discussion with the team about reasonable expectations. Almost anything can be shop-fabricated to the desired level of detailing and finish. This will depend on the fit-out of the fabrication shop and the experience and expertise of the ironworkers, as steel fabrication is still predominantly craft-based. In a shop scenario, increased difficulty will naturally result in increased costs but have little impact on safety concerns with the exception of repetitive strain injuries. These are increasingly able to be accommodated with robotic welding; however, the use of robots should not be considered a panacea as, they are not always capable of access.

Robotic welding would not have been possible for these connections, as there is limited access to the rear of the connection. The spacing of the dense arrangement of tubes needed to account for access to perform the welds. Even then, the ironworkers needed to have rest breaks, as the work was still awkward and highly repetitive.

For erection and site work, in particular when it comes to long spans, cantilevers and working at height, risk must be assessed. Where necessary, either the form of the design or the fabrication and erection methods must be modified to achieve the minimization of risk.

Some basic assessment issues include:

- mandatory personal protection protocol for all workers, regardless of the country of work. North American and European safety standards are very high and when work is done in less safety-driven countries, contemplate modifications to the project details to create a safer work environment; this could in part be:
- elimination of welding at height, where the location is exposed (long-span or cantilever situation)
- innovative means to pre-assemble larger elements on site to lift as a whole, avoiding many hours of at height connections

PUENTE DE LUZ
TORONTO, ON, CANADA, 2011

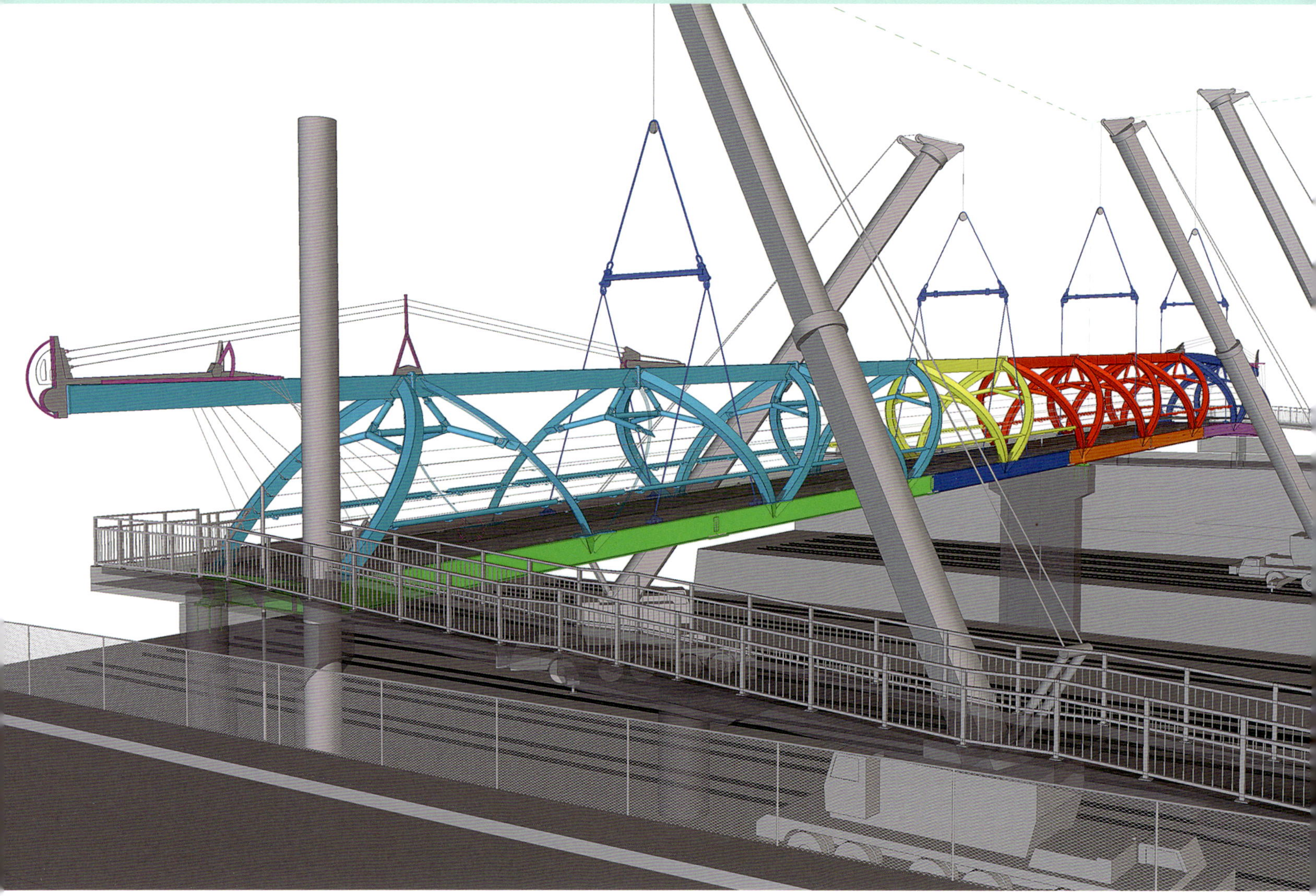

The Puente de Luz pedestrian bridge was to be erected over multiple rail lines that carry all train traffic in and out of the primary train station. This Tekla Structures image shows the four mobile cranes that were used for a coordinated lift during a five-hour night closure. The bridge segments were of varying lengths as determined by the location of the cranes such that the loads were balanced.

Pedestrian bridges have become important urban features. Steel is a particularly appropriate choice of material, as it enables quick erection times with minor disruptions to the traffic flow. In cases where it is not physically possible to lift a pedestrian bridge as an entire unit, multiple cranes can be used to facilitate a simultaneous lift. This must be done quite precisely, so that the splicing operations can be carried out while the sections are still attached to the cranes. In this way, when the cranes are removed, there is no need for a temporary support system that would interfere with the traffic below. This practice is essential when erecting bridges over train tracks or roadways that cannot be closed for long periods of time and where temporary support systems cannot be accommodated.

DESIGNER:
FRANCISCO GAZITÚA

ENGINEER:
PETER SHEFFIELD

STEEL FABRICATION AND ERECTION:
WALTERS GROUP

As the bridge provides access over a major rail corridor, a galvanized metal screen was used to deter suicide attempts. The mesh is fine enough to prevent climbing yet transparent enough to allow for views towards the city.

The splice connections for the bridge were planned as simple bolted connections, which worked well with the aesthetic of the overall connection strategy for the bridge. One of the splice connections is visible towards the bottom left on the deck support member.

A closer view of the overall strategy of bolted connections. This allowed for a maximum share of shop fabrication of the major curved components and an expedient on-site assembly of the four major bridge sections assembled during the night lift.

A detail of the connection between the galvanized rope, the galvanized steel grid and the structural steel of the bridge. Stainless steel clips were used to attach the pieces for precision and durability. Alignment of the various elements was critical to the clean detailing of the barrier.

JOHN STREET PEDESTRIAN BRIDGE
TORONTO, ON, CANADA, 2015

The splice connections located under the bridge and joining the two sides were comprised of simple bolted end plates. These will not be highly visible in the completed location of the bridge so did not have highly specialized requirements.

It is preferable to install a bridge using the minimum number of cranes. Mobile crawler cranes are normally chosen for this purpose, as they can be brought into and out of the site expediently. The crane capacity will be determined by the tonnage of the lift, combined with the required arm extension to prevent tipping. It is important that the surface on which the crane will sit during the lift is adequately firm. Care is taken to monitor rainfall and site conditions.

The John Street Bridge was transported to the site in sections and assembled in the precise location specified for the best positioning of the crane to perform the lift.

ARCHITECTS:
DTAH

ENGINEERS:
HARBOURSIDE ENGINEERING CONSULTANTS

STEEL FABRICATION AND ERECTION:
WALTERS GROUP

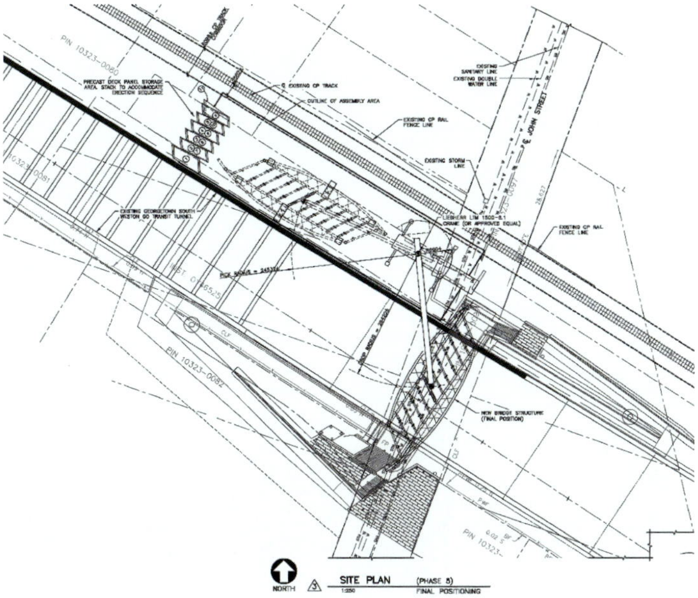

The AESS pedestrian bridge that was to be installed over train tracks required careful coordination and positioning of the mobile crane to pick up and swing the bridge into position. The elements of the bridge arrived to site and were assembled in a staging area that was fortunately available between live surface tracks and recessed tracks in the cut beside.

Temporary structures (in orange) were created to provide temporary support for the eccentrically shaped curved bridge elements as they were assembled in the staging area.

The crane was positioned so that its arm could reach to pick up the completed bridge and swing it into position. The lift took place at night, as it was the only time that the live tracks could be shut down. The tracks precluded the use of shoring or temporary support systems, requiring the erection of a fully completed element. While the fit-outs of the railings and walkway were done after the primary frame was in place, the stainless steel tension system tying the curved arch and the base structure was in place to prevent deformations during the lifting processes.

WEMBLEY ARCH

WEMBLEY NATIONAL STADIUM, LONDON, UK, 2007

The tubular arch at Wembley Stadium is permanently fixed into its inclined position with the aid of a series of forestay and backstay cables. The arch is stationary but is central to the support of an operable roof system to allow daylight into the stadium.

ARCHITECTS:
FOSTER + PARTNERS

ENGINEERS:
MOTT STADIUM CONSORTIUM

STEEL FABRICATION AND ERECTION:
CLEVELAND BRIDGE, HOLLANDIA

One of the biggest challenges of steel erection is making the dreams come true that are easily conjured in the digital world, a place where gravity does not exist. The limits of construction are often pushed when it comes to long-span structures, most particularly where the situation is exacerbated by extreme height and the inability to position temporary shoring towers or permanent supports. The arch over Wembley Stadium is one such project. The arch serves to support a retractable roof and was designed so that no columns are needed to assist in its support, meaning clear paths of view for all in the stadium. The arch has a lattice form, consisting of 41 steel rings that act as diaphragms, connected by spiraling tubular chords. The arch is 7.4m/24.3ft in diameter at the base and weighs 1,750 tonnes. The tapered support hinges weigh 70 tonnes and are in turn supported on concrete bases with 35m/115ft deep piles. The arch structure is 133m/436ft in height with a span of 315m/1,033ft, making it the longest single-span roof structure in the world at the time of construction.

The design called for a fully welded assembly. The welding of the basic diaphragm rings and the stub tubes forming an X at the connecting point took place in the shop, as the requirements were more exacting. Larger units consisting of three rings and the connecting tubes were fabricated on site. Jigs were set up on site to permit the welding operations to take place in temporary sheltered structures erected for this purpose. The fabrication took approximately 10 months. Given its immense size, the entire arch had to be assembled in the correct position for unified lifting on site. It was not possible to erect it in sections, as shoring towers were not an option.

The challenge was then to arrange to lift the fully assembled element into place in a fashion that would not distort and by a means that could handle the extraordinary load. The arch was winched and held in place by five restraining cables. The rotation of the arch from its position on the ground to its final incline of 112° took place in stages to allow for adjustments to the temporary cabling support systems. In its permanent position the arch is held in position by a series of forestay and backstay cables tied to the main stadium structure.

A view inside one of the temporary fabrication sheds on site. Temporary supports were constructed to hold the base rings in place and to permit the welding of the joining tubes. Priming and paint was held back from the portions to be welded.

The hollow tubular steel frame was designed as a fully welded system, including all site connections. For the integrity and durability of the structure over time, the all-welded solution was preferred over bolted connections, which may have been more vulnerable to oxidation given the wet weather experienced in the location. Welding could fully seal the tubular structure from water penetration. For regular access to the lighting system on the arch, and also to facilitate inspections, a permanent access system runs up through the arch.

Prosteel, Tekla and CAD were used for the design development and essential to the positive outcomes of this challenging project. The entire lifting process was computer controlled to ensure accuracy and timing given the need for precise coordination in order to prevent uneven lifting that would have undermined the process.

Logistically the arch assembly at grade and lift had to be completed ahead of the construction of a large section of the stadium in order to maintain access for the winches, lifts and cranes during the site fabrication and erection process. This sequencing had a large impact on the timeline of the project, as much work was unable to proceed until the arch was securely in place.

None of the welds on the arch have been remediated. The sheer scale of the structure, combined with an understanding of the impact of distance to view, led to this economical decision. The multiplication factor also played into the decision to forego remediation.

A view of the connection between the arch and the concrete foundation. A ball joint allowed for movement in multiple directions when setting up the arch position. Anti-climbing materials have been applied after the fact by the owners as a deterrent. Raising the height of the concrete base may have been helpful in offsetting the urge to climb, which is an unfortunate inevitability on many projects.

A view of the conical concrete foundation and the tapered arch support, giving a better impression of the scale of the arch.

ECONOMICALLY DRIVEN STRATEGIES

CHAPTER 7

As discussed in Chapter 3: Managing Complexity, there are many different ways in which the presentation of complexity in steel design can be pursued, some costlier than others. This chapter will look at a number of strategies that can soften the fabrication requirements for the steel to some extent, while still arriving at an aesthetically high level final state.

1. THE DISTANCE FACTOR

The overall strategy behind the category system for specifying Architecturally Exposed Structural Steel is to allow the distance to view to soften the fabrication requirements for the steel. If a surface or element is situated further than 6m/20ft from view it is unlikely that the viewer will either be able to see or appreciate fastidious detailing. This is considered a 360° distance, and this would hold also for multi-storey atrium spaces, for instance.

The AESS structure of Tokyo Midtown, Tokyo, Japan, by SOM and Nikken Sekkei, 2007, is a great example of the effective application of the distance factor. Different approaches to detailing and systems selection have been strategically applied to the lower level structure, the spanning system and the extreme high level structures for the canopy. Layering effectively obscures the view to the most distant steel.

The distance factor is of significance in particular when it comes to weld remediation. Although AESS 3 projects are permitted to ask for all-welded structures – including connections between the smaller transported elements as well as internal splice connections – it is permitted to use hidden or discreet connections if deemed appropriate based on viewing distance. Hidden or discreet connections may also come into play where it is physically difficult to provide proper access to complete site welds, including remediation tasks.

In larger projects, such as airports, where the number of connections is extremely high, the multiplication factor can be considered when contemplating the relative importance of weld remediation to the project. It may be deemed not worth the expense in terms of an aesthetic benefit to cost ratio to perform the remediation.

TOKYO MIDTOWN

TOKYO, JAPAN, 2007

The complex steel structure at Tokyo Midtown applies a hierarchical system of detailing. The project incorporates two types of curvature – bending at the connection of the steel tree branches to the trunks, and faceting to use straight segments to create the dual curvature of the high-level trusses.

ARCHITECTS AND ENGINEERS: SKIDMORE, OWINGS & MERRILL LLP WITH NIKKEN SEKKEI

The large AESS canopy that covers the connecting plaza between the Tokyo Midtown towers makes use of the distance factor to create a hierarchy of steel fabrication and connection types. There is a clear separation of the detailing languages of three unique systems. A space truss system was used to span between the custom-fabricated bases of the tree columns and their immediate branches. Another subsystem of tensile bracing within the truss provides rigidity and adds further visual layering and complexity. The skylight system is supported by a simpler system of wide-flange (Universal) members finished with a darker grey to allow this system to recede even further from view.

The degree of weld remediation has been aligned with distance to view. Only the tree structures are fully welded, with weld remediation limited to key splice connections in the round trunks and rib plates. The types of details have been selected as a function of the different systems, yet there is an overall coherence.

This night view of the roof structure illustrates the curvature of the truss elements. These have been created using straight tubular members, connected with spaceframe-type nodes that easily accommodate varying angles, and braced using tensile bracing. The lighting highlights the fabrication detailing more than the daylight condition, as the backlighting during the day tends to suppress the reading of the details.

A high level of weld remediation was used on the trees, both on the splice welds that attach the tree branches to the trunk and on the butt welds on the plates that form the ribs. The connections between the plates and the tubes use simple, unremediated fillet welds. At some of the splice locations, evidence remains that not everything could be remediated, but considering the distance to view, the work is more than satisfactory.

A close view of the upper-level truss. A custom conical element that includes a round nodal connection has been welded to the top of the branches. The temporary bolt tabs at the tops of the branches have been cut tightly to the tubes, but not remediated because of the distance to view. Simple discreet bolted connections create the connections for the vertical chords of the truss. The plates extend outside of the tubes to permit fillet welding. Standard tensile connectors have been used to brace the two-way truss. Dark grey painted steel supports the skylight. Standard orthogonal connection details were used at the skylight level because of the distance to view.

SHENZHEN CULTURAL CENTER
SHENZHEN, CHINA, 1998–2008

The Shenzhen Cultural Center is composed of a concert hall and a public library. Although physically separated, the buildings share a common language of detailing. Each has two complex features – a large, irregularly angled skylight over the entrance/atrium space and an undulating glazed wall alongside the adjacent plaza. The detailing of these elements is quite distinct from each other. The undulating façade is supported by a set of exterior stainless-steel-clad fins. This provides shade to the glass and leaves the interior glazed surface free from texture, although the curtain wall is framed in a diamond pattern to complement the angularity of the skylights. The textural component in the interior is the highly sculptured structural steel support system for the glazed roof. It features specialized detailing for the tree-like supports and a less visually prominent system to support the glazing, following the distance-to-view logic.

The setting of hierarchies based on nearness to view allows for simple discreet connection detailing at the high skylight level. The lower support steel is completely clad and could therefore be designed as standard structural steel, avoiding any AESS type designation.

ARCHITECT:
ARATA ISOZAKI

The undulating glazed façade of the library to the left can be seen against the mountainous glazed roof system atop the concert hall. The two systems are completely separated in terms of detailing, although from a material point of view, they complement each other.

The angular skylight forms of the Shenzhen Cultural Center are supported by a tree-like system that is structurally distinct from the support system for the wall glazing. This provides an interesting opportunity to create a different focus on the detailing of the resulting "tree-like" roof supports. Rather than exposing the steel in the concert hall lobby, the lower two-thirds of the structure has been concealed with a ribbed gold-colored cladding system.

The gold-colored ribbed cladding on the tree-like support system stops short of the welded connections at the skylight level. Round tubular shapes define the overall peaks and valleys of the skylight, as the odd geometries could be better accommodated by nodal connections. The glazing is supported by a purlin system fabricated from more standard wide-flange (Universal) sections.

Round tubular steel accommodates complex nodal connections. Wide-flange (Universal) members support the smaller skylight framing module at the highest level. These members are more economically detailed, given the extreme distance from viewing level. The plates cut into the tubes protrude so that fillet welding could be used. Very simple, discreet bolted connections using plates splice the larger, shop-fabricated tubular element sections and attach the purlin members, which was entirely done on site, simplifying the erection process. The shape of the skylight does not allow to use standard methods of cleaning. Window washers need to abseil the glazed slopes to perform this task, so that the glazing must be designed for this type of loading. The view from the lobby of the library shows that the color of the decorative member cladding is changed to silver/white in order to differentiate the reading of the space from the concert hall.

BARAJAS AIRPORT
MADRID, SPAIN, 2005

An overall view of the Barajas departures hall showing the undulating roof structure, with only its bottom flange exposed, supported by tapered steel columns.

ARCHITECTS:
ROGERS STIRK HARBOUR + PARTNERS AND ESTUDIO LAMELA

ENGINEERS:
ANTHONY HUNT ASSOCIATES,
TPS WITH OTEP, HCA

The brightly colored Architecturally Exposed Structural Steel at the Barajas Airport in Madrid creates the signature aesthetic of the project. Given the sheer size of the facility, it was prudent to look at the distance factor in conjunction with the multiplication factor when deciding on the approach to detailing. The project used a number of specialized fabrication methods for steel products that contribute to the high level of complexity. These include angled tapered steel columns and a large undulating roof support system.

The angled conical tubular legs could have required a fastidious approach to weld remediation, given that the view was potentially up close. Cones are created using brake forming, so that there are strict limits of the sizes of plates that fit into the brake press, necessitating internal splice connections. In order to maintain visual continuity in the long supports, an all-welded joint was essential. However, the selected yellow color masks any shadow lines that might result from the protrusion of the weld above the adjacent surfaces, so that these have been left unremediated.

The undulating curved steel used to support the roof is hidden from view by the suspended wood ceiling, leaving only the bottom chord of the support exposed. Therefore, the detailing of the support system could be quite standard, focusing on a higher level of finish only on the bottom flange.

The connections between the vertical supports and the roof structure use nicely detailed bolted connections. They are relatively discreet and allowed for a faster erection process.

The base connection of the tapered columns sits slightly above eye level, allowing for a fairly straightforward bolted connection. The connecting segments of the cones have been butt-welded and the welds left unremediated, as the color and lighting downplay the detail quite naturally.

A pair of bent elliptical tubes has been used to create this Y-shaped support in the baggage area.

A custom bolted connection is used to join the Y-shaped support to the underside of the large curved roof members. The connecting piece is beautifully shaped and comes to the site as part of the horizontal member that joins the top ends of the Y to keep it from deflecting while waiting for the attachment of the roof system. Eventually this constitutes a permanent bracing system. All of the welding can be completed at the shop, leaving only bolting exercises to be performed on site.

The splice connection at the point where the branches meet the vertical section of the Y-shaped column have unremediated welds. Evidence can be seen of the removal of the temporary connecting plates. These members would have been far too large to ship as one piece, so that on site welding was required. Given the height of the weld above the natural viewing position, it was logical to require only good-quality welding and not grind the connections.

2. SEMI-EXPOSURE AND THE USE OF SCREENED ELEMENTS

The idea of layering via the use of screens or semi-transparent membranes located in front of the structural system is an approach that can achieve a unique presentation of complex steel. The structure is viewable through the screen, albeit often in near silhouette. The fabrication and material requirements for the screen and the structure can be clearly separated. The limited exposure of the steel is able to reduce the AESS fabrication requirements. Fabrication requirements will be different for interior versus exterior conditions in terms of durability, weathering, fire protection and finishes.

Two primary conditions are prevalent in this type of use: in one, the structural steel is in support of a double façade envelope; in the other, an interior finish system for ceilings and walls is suspended in front of the steel to permit a limited view. In both cases, requirements for the passage of daylight will impact the level of transparency of the screen element and in consequence the level of view of the steel. The less the steel is seen, the less refined the fabrication detail can be.

The screen elements are typically non-structural and can therefore be fabricated from thinner (plate) materials. Durable materials such as aluminum, stainless steel, glass and membranes can and should be used for weather resistance. Galvanized sheet products tend to be less durable, as they tend to be quite thin. Galvanizing requires steel thickness with enough depth for the sacrificial nature of the zinc coating to work. If plate materials are drilled or decorative holes are cut prior to hot-dipping, the heat of the galvanizing bath can cause undesirable deformations. Where numerous holes are drilled to facilitate connection to the supporting structure, the galvanized coating can easily be damaged by mechanical fasteners such as screws and bolts, resulting in rapid deterioration and oxidation. Durable metals are preferable, as they have uniform material properties throughout and therefore do not have corrosion-related issues with cutting and drilling. Care must be taken when using different metals to avoid interaction that may cause galvanic corrosion. Care has to be taken that the natural oxidation of materials, such as copper or weathering steel, will not result in run-off and staining.

A variety of materials can be used as interior screens, as durability is not an issue, although access for cleaning and maintenance remains a concern. As the screen creates a load on the steel structure, minimization of weight is a factor. Different methods can be used to stiffen thinner metals that are intended to be used in applications where the visual access is via holes in the material rather than in spaces between panels, as in the case of slatted systems. In most instances, they will be fitted to a system of smaller metal frames that needs a method of attachment to the main structure. If these are exterior, the frames will normally be fabricated from galvanized structural steel for durability, unless stainless steel can be afforded.

Precise CNC methods for cutting the sheet materials are available, which can succeed in creating very fine detailing in the overall appearance, even with fairly simple strategies.

The detailing of the primary support structure which sits behind the screen can be substantially softened. Even if in partial view, there is little point to fabricating beyond AESS 2. This will preclude extensive on-site welding and favor more simple bolted connections. In exterior applications, the attachment systems must be designed for corrosion resistance and to preclude places for water and snow to puddle. Clean lines in the overall finished form become all the more important.

It will be necessary when designing the screened system, acknowledging that the steel will still require periodic cleaning and maintenance, to ensure that there is proper access for these operations. The space around the steel must be large enough to accommodate physical access and cleaning equipment. In double façade systems this usually infers the inclusion of permanent walkways that in many instances will double as shading devices.

The slatted wood screen that clads the ceiling of the Barajas Airport is relatively tightly spaced, barring much of the view into the structural space beyond. The daylighting is effected not through the screen but in restricted locations through the roof. Only the bottom flange of the curved steel beam is left exposed.

The stainless steel-screened façade on the Cooper Union, New York City, NY, USA, designed by Morphosis, 2009, uses perforations to provide a very limited, almost silhouette-type view of the steel beyond. The thin screens are supported by a system of steel frames attached back to the primary system. Galvanizing has been used to protect the structural steel.

A glazed and partially fritted glass façade allows for an obscured view of the chaotic galvanized steel frame at Federation Square in Melbourne, Australia, designed by LAB Architecture and Batesmart, 2002. This arrangement also affords some environmental protection for the steel, which, in contrast, is fully exposed on the interior of the building courtyard.

Elsewhere on Federation Square, triangulated panels of perforated screens and opaque materials interplay with an unobstructed view of the steel grid beyond. The system provides overall screening to the relatively conventional curtain wall system behind. Although the effect may appear chaotic, a prefabricated modular system based on rotated triangles was used. This reflects some limitations of digital design and steel fabrication methods of the period.

The ETFE membrane at the Watercube in Beijing, China, designed by PTW Architects, 2008, partially obscures the view of the complex tubular steel system beyond. The membrane encloses a double façade system and maintains pressurization. The overall form was digitally carved out of Weaire-Phelan Foam.

The exterior view of the ETFE membrane on the Watercube provides a much more limited view of the steel structure beyond. The relative transparency of membranes, interior versus exterior views, coloration and integration of thin-film photovoltaics can all be used to manipulate the level of visibility and reading of the visual complexity of the steel.

BEIJING CAPITAL AIRPORT
BEIJING, CHINA, 2005

ARCHITECTS:
FOSTER + PARTNERS

ENGINEERS:
ARUP

The brightly colored spaceframe of the Beijing Airport can be seen through the slatted screen. The effect provides an impression of the steel, as some of the members can be seen where illuminated by daylight, but masks most of the connection details.

The use of a slatted screen has been an effective means to partially conceal the view through to the spaceframe structure that comprises the large spanning curved roof plane at the Beijing Airport. This sort of ceiling detail has not been uncommon in many transit-type projects. Although the steel framing system for the roof could be classed as architecturally exposed, the detailing is able to be greatly downplayed, bringing some economy to the project over a more overtly expressed system. Still, the geometry has been carefully planned and coordinated with the triangular skylight locations as well as with the subtle triangulated pattern of spacing joints that has been used on the ceiling. The screen also assists with concealing all of the mechanical systems that reside within the truss and permits air flow though. The use of an overall brightly painted color provides a glow behind the screen and ultimately also helps in concealing the connection details and contained systems.

The use of the screen allows for a clear change of approach to detailing for the highly expressed AESS windbracing system that supports the curtain wall system. The diamond-shaped cuts in the lattice add visual interest and complement the placement of the triangulated skylights. They also allow for the removal of panels for maintenance access.

The slats in the ceiling have been spaced far enough apart to create a dynamic and deep feel to the space between the suspended ceiling and the high skylight level. The skylights have been used to illuminate the steel, creating interesting areas of bright color and shadow. The uniform color of the structure and the mechanical systems allows increased freedom in the detailing and logistical planning of this zone.

A view into the connection between the spaceframe and the more finely detailed windbracing system. The abrupt change in the structural systems and the associated levels of detailing are hidden. The constant dimensional changes along the curved outer edge of the roof are dealt with via small pinned connectors.

The triangular, fully welded trusses are used to support the fully glazed façade of the lounge area. The outward lean is being restrained by the connections to the spaceframe structure, as seen to the left.

SHENZHEN BAO'AN INTERNATIONAL AIRPORT
SHENZHEN, CHINA, 2013

A metal screen with a painted finish blocks a direct view to the spaceframe system at the Shenzhen International Airport. The color is able to reflect light back into the space at night as well as filter direct solar penetration during the daylight hours. A more standard fully welded HSS system has been used to support the outwardly sloped curtain wall system.

ARCHITECTS:
STUDIO FUKSAS
WITH BIAD

ENGINEERS:
KNIPPERS HELBIG
ENGINEERING

A metal screen partially blocks the spaceframe truss from view, which handles the long spans between the column supports. The complexity of the varying shape of the terminal's cross section, including large areas of curvature, is accommodated by the flexibility of the spaceframe system. The screen elements permit a partial view of the exposed steel structure, greatly softening the need for a high level of fabrication detailing. The screen is instrumental in assisting to block sunlight penetration from the rooftop skylights into the interior. At night the screen succeeds in bouncing illumination back into the space. As is customary with this sort of screen application, the service systems that are housed in the truss space have been painted to match, allowing them to effectively disappear.

The geometric forms of the thin metal screen have been designed modularly to easily adapt to the changing curves of the interior form, creating an increased vitality.

A view up the sloped curtain wall bracing system. A Vierendeel truss fabricated from all-welded square HSS members creates a simple visual contrast with the hexagonal shapes of the screen above.

The modular metal screen has the added advantage of masking the view to the services in the ceiling space, softening the potential design for view requirements of these services. As durability is not an issue for the interior application, there was no need to use aluminum or stainless steel products. The bent form of the metal helps to improve its rigidity, allowing for the use of a thinner, lighter product, thereby reducing its weight on the structure. Lights within the truss space add an interesting, dynamic level of illumination at night.

LOUVRE ABU DHABI
ABU DHABI, UAE, 2018

The design for the 180m/590ft diameter dome, which covers the conditioned museum spaces and open courtyards at the Louvre in Abu Dhabi, used the strategy of layered screens, reflecting the cultural form of the mashrabiya sun shade, providing semi-conditioning for the outside spaces. The layering of the interior and exterior skins creates a screened view of the 5m-high steel frame support system, putting the focus on the detailing of the screen rather than the steel truss system sandwiched in the middle.

The four exterior layers of the dome were fabricated from stainless steel, whose superior corrosion resistance responds to the marine location. The four inner layers were fabricated from aluminum, as a means to decrease the overall weight of the structure and acknowledging the more protected location in terms of weathering.

The steel frame for the dome was fabricated from 10,000 structural components that were preassembled into 85 elements, each weighing up to 50 tonnes. The sizing was to work towards optimum values for transportation and erection.

ARCHITECTS:
ATELIERS JEAN NOUVEL WITH WAAGNER BIRO

ENGINEERS:
BURO HAPPOLD

The large lattice-covered steel dome atop the pavilions is typically only visible from a distance.

The details of the steel truss are visible through openings in the screen. The square HSS tubes have a simple bolted plate connection to attach to the specially fabricated hexagonal nodes. The plates protrude from the tubes, permitting simple fillet welds. It is the overall shape and form of the truss rather than the detailing that makes the impression from below. HSS members over wide-flange (Universal) shapes work well with the complexity of the angles of the truss. They also prevent surfaces where water could collect on the truss exposed to the environment.

The multi-layered aluminum screens obscure the view from below to the primary steel support truss. The overall color palette assists in blurring the detailing, as well as being a good color choice for weathering in this dusty climate.

Multiple layers of stainless steel form the roof covering for the dome. The layering amplifies the dynamic variations more subtly than a single layer with perforations could have. Layering also permits variability based on orientation, a strategy used by Nouvel also for the Doha Tower.

During the hot daytime hours the multiple layers of the screen provide a high level of solar control, while simultaneously allowing for ventilation and dappled lighting.

CALTRANS DISTRICT 7 HEADQUARTERS

LOS ANGELES, CA, USA, 2005

A fine aluminum screen covers the galvanized steel frame, providing shading for the façade. The silhouette of the steel can be seen and in particular places the screen is removed to completely expose the rugged steel frame. Here the system is manipulated to create an oversized street number as iconic signage for the building.

ARCHITECTS:
MORPHOSIS

ENGINEERS:
JOHN A. MARTIN & ASSOCIATES

The Caltrans District 7 Headquarters was designed with a double façade envelope, comprised of a perforated aluminum skin mounted on a framework of galvanized Architecturally Exposed Structural Steel. The semi-concealed structural steel took on complex geometries to support the angled planes of the perforated façade, allowing prominent exposure of the façade screen.

The aluminum shading screen is very thin and light and provides a better corrosion resistance than a galvanized sheet steel solution. Operable panels of screen are located in front of the glazed portions of the climate-controlled areas of the envelope, permitting user control of direct sunlight and view.

The connection language of the steel frame uses standard endplates and bolted tubular connections. Due to the screen, geometrically tight discreet connections were not required; however, clean lines and an organized approach to the structure were important to the aesthetics. HSS rather than hot-rolled steel has been used in the most visible areas to achieve a cleaner look.

Exposed frames need to be detailed in such a way that no rainwater or snow can accumulate. The use of HSS sections versus extruded shapes, such as wide-flange sections, is a means to decrease places that could allow for water accumulation. Although the geometric conditions of the frame change around the building, the connection detailing has been kept unified.

The silhouette of the galvanized frame appears behind the mask of the aluminum screen with its varying levels of perforation. Daylight penetration varies according to the conditions of the curtain wall behind. In this way, the detailing could place more attention on the technically important aspects of fabrication and durability. The tower itself is orthogonal. The façade form has been used to create the angular look of the building.

A view up inside the double façade showing the use of various detailing approaches to the galvanized steel framing. Access walkways allow for ease of cleaning and maintenance of the glazed façade as well as the screen. As this view is not easily apparent from the pedestrian perspective, the steel is allowed to visually recede in the design.

SAN FRANCISCO FEDERAL BUILDING
SAN FRANCISCO, CA, USA, 2007

A perforated stainless steel façade, supported by galvanized steel, creates the angular Deconstructivist geometries of the San Francisco Federal Building. Roof elements and accessory structures add to the exaggerated angular palette.

ARCHITECTS:
MORPHOSIS

ENGINEERS:
OVE ARUP &
PARTNERS

A perforated second skin was critical to the intention to make the San Francisco Federal Building a naturally ventilated office space. The skin provides natural shading to the curtain wall. A specialized steel system allows for a spatial separation of the screen and the curtain wall. Significant portions of the structural system that supports the stainless steel screen have been fabricated from galvanized steel. The predominant use of tubular members avoids puddling traps.

The unusual geometry of the tower is created by the screen element, while the tower block itself is regularly shaped for economy of construction. Roof elements, shades and canopies on accessory structures provide extreme angles in tune with the Deconstructivist style.

A custom-fabricated bent column supports the end of the shading canopy. The grey color of the galvanized steel and its technical, wide-flange appearance complement the durable stainless steel material used for the canopy.

The support structure of the rooftop shading screen is fabricated from galvanized HSS and wide-flange (Universal) members. The overall aim was to lend a highly technical look to the steel. On-site connections make predominant use of bolting.

A discreet tube-to-tube connection uses a sleeve insert to create a trimmer appearance than would be possible via the typical bolted end-plate connection. The stainless steel screens wrap around a lightweight frame that provides sufficient strength to permit the panels to cantilever from their galvanized HSS purlin supports. This recognizes the absence of snow loads in this climate zone.

A variety of member types has been used for this canopy, each to fit with the geometry and purpose of the structural element. The cantilevered beam has been tapered to reflect its function as a moment-resisting element. A tapered beam is a custom-fabricated piece. The inverted pyramid that picks up the load from the spanning beams that support the purlins, which in turn support the screens, has been fabricated from round HSS members. This works more easily with the geometry of these connections. This all-welded member has been site bolted to the beam and has bolted pin connectors at its end points.

3. MAKING THE STEEL RECEDE

Although the use of complex structural steel may well be the basis for the form of the project, it may not always be placed visually in the forefront of the design. Some projects will choose to fully clad the steel. This may be done due to weathering issues, fire protection constraints or for aesthetic reasons. Other projects may partially expose the steel as a support system but draw attention towards the cladding system and away from the steel itself, effectively making it recede from view. In both cases the steel is the basis of the form of the design but not the primary material focus.

Even though the steel is not visually featured, the detailing will be equally critical to the project outcomes. Where the steel is to be clad but its form still expressive, the connection design will be important as the means to keep the bulk down, and so standard framed connections might be unsuitable, instead preferring connections similar to discreet bolted types. Where the steel is exposed, the detailing may need to be downplayed to reduce visual attention to the structure and let it serve a supporting role.

The steel of the chaotic structure at the Jewish Museum Courtyard in Berlin, Germany, designed by Daniel Libeskind, 1999, is clad. Although the fabrication detailing is completely concealed, the essence of the steel is still felt in the space. Nevertheless, it is important to maintain a slender appearance of the members and use very constrained connections, similar in design to discreet connections.

These two photographs of the Addition to the Royal Ontario Museum, Toronto, ON, Canada, designed by Daniel Libeskind, 2006, have been taken from approximately the same vantage point. The presence of the structure is evident in the finished space, highlighted by the glazing pattern, although the steel is completely clad in gypsum board. The cladding over the wide-flange (Universal) sections is quite trim, while the bulkier connections are located behind the ceiling cladding.

In the Guggenheim Museum, Bilbao, Spain, designed by Frank Gehry, 1997, varying approaches have been taken to the aesthetic use of the complex steel structure. Some of the steel is directly exposed. Significant glazed elements cause the steel to recede from view. Although the extensive drywall covering fully conceals the steel frame, its presence is still clearly felt.

The diagrid structure of the Denver Art Museum, CO, USA, designed by Daniel Libeskind, 2006, is completely clad in gypsum board. This can be attributed to the functional purpose of the space as well as to fire protection. The presence of the steel is still felt.

In a similar vein to the idea of using screens, adjacency maybe used to shift the focus away from exposed supporting steel, thereby softening its fabrication detailing. The eye of the viewer will be drawn to a more finely detailed and polished part of a project to the point where the structural support system, though clearly visible, takes a visual back seat. This is different from the "distance to view" strategy, as the steel may actually be very near to view and touch.

THE VESSEL
NEW YORK CITY, NY, USA, 2019

The color palette of the interior blends the stone and steel finishes to downplay the visibility of the steel. The steel supports tend to be hidden in the shadows, also making them recede from view. This allows the design to soften the fabrication requirements of the steel structure.

The large sculptural staircase at the Hudson Yards in New York City has become a tourist destination. It is part of the experience to appreciate all aspects of the structure at very close range. The structural steel of the support system for the mirrored finish cladding has not been given a similar "glossy treatment." Its welds have been neatly done and left unremediated. Hidden bolts or discreet bolted connections provide an overall welded appearance.

The 75 interconnecting dog bone – shaped modules of the staircase sculpture were prefabricated by Cimolai S.p.A. in Italy and shipped to the site by barge. The fabricator preassembled the entire structure prior to shipping to verify its fit. The steel was also prefinished, partly to make the site assembly go faster, but also to protect the steel from corrosion en route. The ends of the large hollow sections were temporarily covered with plywood to keep rain from puddling on the uncoated interior.

The erection was planned to preclude the use of temporary support systems. The elements needed to be assembled in such a way as to maintain a physical balance, a challenge exacerbated by the very tight base dimension of the Vessel and the very large overall dimension across the top. The overall shape cantilevers out increasingly at each level.

ARCHITECTS:
HEATHERWICK STUDIO

ENGINEERS:
THORNTON TOMASETTI

The steel modules of the stair were designed to be fully self-supporting as the erection progressed, so that there was no need for temporary bracing. The on-site connections were all bolted, facilitating a faster erection sequence. The polished cladding of the upper modules was pre-installed, so that there was no need to scaffold the exterior. In front are some of the infill elements, whose support framing was also fabricated from plate steel.

A view of the steel during construction. The bolted interior connections between the modules will all be hidden. The bolting operations were accessed from discreetly located access panels.

To the left, surface-mounted plate steel has been used to complete the plate-to-plate connections, and round-headed TC bolts create a more finished (almost historically riveted) appearance. The engineering took many fabrication clues from the ship-building industry. To the right is the connection between the two modules, with the bolts completely hidden.

Some brake-forming has been used to create curvature, but the majority of the shapes have been created by cutting plate steel to shape and shop-welding it together. The welds have been left unremediated as part of the technical look. The stark contrast with the shiny cladding will leave most visitors fairly unconscious of its presence.

The structural steel support system for the Vessel is not visible from the exterior. The eye is drawn to the reflections on the brilliantly shiny copper-colored cladding, which was attached to the modules prior to being lifted into position, significantly saving on construction time and logistics.

LOU RUVO BRAIN INSTITUTE
LAS VEGAS, NV, USA, 2010

The dominant exterior view of the Lou Ruvo Brain Institute from the adjacent roadway presents an undulating curved façade. It is clad with small panels of flexible stainless steel, which could be custom-fit to the curved structure during installation. The panel sizes are slightly smaller than those on some of Gehry's large projects because of the frequency of the window openings.

ARCHITECT:
FRANK GEHRY

ENGINEERS:
WSP

There are many viewing angles of the Lou Ruvo Brain Institute, and Frank Gehry used a different approach to each to suit the nature of the exposure. This had an impact on the steel detailing. Mostly the steel is seen in a physical and aesthetic support role, and is not the feature. Therefore, the detailing, member selection and connection design could be quite softened. This approach took lessons from the Jay Pritzker Pavilion in Chicago, while the overall façade of the Brain Institute is much lighter and impacted by the presence of 199 openings. Virtually every steel connection was individually designed and fabricated as a result of the unique geometries, and the use of BIM was essential to the success of the project.

A view up the underside of the steel framing that supports the stainless steel cladding of the exterior canopy. Wide-flange (Universal) members have been used for the more vertical supports. Tubular members negotiate the connections between. The angular steel framing is in contrast with the curved exterior beyond. This is another instance of "faking the curve", as discussed on p. 92.

The ruggedly detailed steel framing of the exterior courtyard canopy is revealed under the undulating form, making a purposeful contrast with the stainless steel skin beyond.

An approximation of the curve in the use of steel: smaller pinned supports mediate between the angular shape and the prefabricated panels, which more closely form the actual curve of the cladding.

Plate steel is used for most of the curved forms where it is necessary to fit to the junction of the canopy. Plate steel is easy to cut to accurate curved forms. The balance is created from triangular or trapezoidal pieces. The larger prefabricated sections have been bolted together on site. Welding operations on the elements were completed in the fabrication shop.

In the interior, the steel is completely clad in gypsum board, in part to satisfy the aesthetic requirements of the rental space and also to provide fire protection. The rugged underside of the steel canopy is viewable through the large glazed opening, creating a purposeful juxtaposition of textures.

4. FAKING THE CURVE

Bending and curving steel remains a largely hand-crafted process, adding to the cost of the contract. Steel rolling and bending must be done at a specialty fabrication shop. Roller-bender facilities are not as common as structural steel fabricators.

Advances in software have enabled a significant amount of curvature in contemporary design. However, what is simple to create digitally remains one of the biggest challenges in steel fabrication. Finding an approach that allows for a curved form does not necessarily entail bending or curving the steel. *(Chapter 8: Curved Geometries* will examine more closely several projects that use true curvature.) It may be possible to use straight segments of structural steel to achieve an impactful impression of curvature. The overall scale of the form, surface or structure may be large enough or distant enough to effectively use straight members to create segmented or approximated curves.

Genuine curvature may be inadvisable with a view to the compatibility of the structure with the façade materials. Where thinner metal cladding, like the stainless steel and titanium used in many projects by Frank Gehry, is quite easily adapted during installation to fit curved substrates, most glazed façades will preferably use flat glass in a faceted approach. The scale of the structural steel needs to be coordinated with the glazing support system, as often the glass panel size may be smaller or more finely grained that the overall structural support system. Certain types of cladding, like cut stone, are not available in anything other than flat panels.

The diagrid framing system for the Poly International Plaza – Dawangjing in Beijing, China, designed by Skidmore, Owings & Merrill, 2016, uses straight tubular steel members to create the elliptical form of the tower. For strength as well as fire resistance the frame will be concrete-filled.

Faceting is a common strategy on diagrid structures, as it allows for the use of conventional glazing systems and avoids the need for expensive curved glass products.

The cladding on the Disney Concert Hall in Los Angeles, CA, USA, designed by Frank Gehry, 2003, is flexible and dimensionally small enough to fit to a genuinely curved substrate.

The curved stainless steel façade of the Disney Concert Hall is primarily supported by a structural steel framework comprised of straight segments. A lighter steel framing system mediates between the more angular geometry of the structural steel and the curvature of the cladding. Mesh is used on the exterior portions of the structure to prevent birds from roosting.

The Phoenix University Stadium, Phoenix, AZ, USA, designed by Peter Eisenman, 2006, uses the scale of the building to convey an impression of curvature. The structural steel supporting the horizontal flat cladding panels is fabricated from fairly standard HSS sections. The main frame angles are complemented with a finer network of smaller spanning structures. Viewing distance has allowed the use of fairly standard bolted connections.

The scale of the stadium has allowed for the use of all straight members and flat cladding for its double curvature. The panel sizes have been carefully coordinated with the structural spacing to ensure that the faceting of the façade is not extreme.

CURVED GEOMETRIES

8
CHAPTER

The Osaka Art Gallery, Japan, designed by Cesar Pelli, 2004. The highly sculptural entry pavilion is fabricated from stainless steel. This material required full remediation of splice welds, fastidious workmanship to support its aesthetic demands, and a high level of craft and expertise to achieve the highly irregular curvature.

The incorporation of curvature into steel design is the area with the greatest gap between what can be digitally designed and what can be reasonably fabricated. It requires of the designer a high level of understanding as to limitations on accuracy and how to best work with steel fabricators and a subset specialty fabricator, the roller bender, to realize a curved aesthetic.

Methods of curving steel are covered in detail in Understanding Steel Design, Chapter 8: Curved Steel.

One of the most striking departures in the last 20 years of steel design is the widespread incorporation of curvature. As discussed in *Chapter 2: The Digital Revolution*, the ability of design software to produce smooth curves has fueled the desire for this aesthetic. Prior to the pertaining advances in design, calculation and fabrication techniques, curvature, to a large extent, had been limited to and characteristic of monolithic reinforced concrete construction. The invention of the geodesic dome by Richard Buckminster Fuller, and experimentation with the application of bubble geometries to steel structures by Frei Otto, were long considered unusual, and their extraordinary nature prevented them from being incorporated into mainstream architectural design. Yet even the majority of these structures "faked the curve," as they were comprised of straight members connected via nodes. Their scale allowed the illusion of curvature. These early experiments tended to be based on highly regular and repetitive geometries. This method of creating curvature is still quite valid today and leads to more economical solutions, as referenced in *Chapter 7: Economically Driven Strategies*.

While the exuberant exostructure of the Osaka Art Gallery has been fabricated from stainless steel, which requires fastidious care in detailing and fabrication, the structure immediately behind it, directly supporting the glazed enclosure, has been fabricated from round HSS sections that have had a finish applied that blends with the overall appearance. This affords some savings to the project.

1. FABRICATING CURVED STEEL

A brief review of methods of creating curvature is critical to the discussion of the relationship between curvature and complex steel, as the various methods necessitate quite different approaches to the processes of design and construction.

Primary methods for creating curvature and the respective methods:

- Faking the curve
  - using straight segments to facet the design
  - applicable to large-scale forms
- Bending the steel
  - using a three-point pressure-bending system
  - equipment that can roll structural shapes with even curvature
  - equipment that can roll plate material with even curvature
  - sections or plates that are too thick are difficult to bend
  - curvature manually checked with a template
- Brake-forming
  - use of a brake press to bend plate steel
  - applicable to irregular curvature
  - used for making conical shapes
  - curvature manually checked
- Plate-cutting
  - transfer of digital information to devices to cut plate steel into curved shapes
  - useful for creating larger web members that can be fabricated into large curved beams
  - accurate curvature based on digital files while taking into account minor loss of steel due to the cutting process

Bending the steel:
In a three-point bending process the steel is rolled back and forth between a set of forms to fit the structural shape. This method can be used for HSS, angles, channels and wide-flange (Universal) shapes. To prevent deformation of the shape, the curvature cannot be extreme and the section must not be too heavy nor too light.

Plate-rolling:
A three-point pressure system is used to roll plate.
The equipment size determines the maximum width of the plate.
The curvature is verified by checking it against a template.

Brake-forming:
The steel is carefully marked with lines to indicate the required pressure points when the steel is tapped in the brake press. This requires a very skilled operator to ensure that the pressure is not so heavy as to translate the lines through to the exposed face of the steel, as well as to decide on the layout of the pressure lines and the degree of impact from the brake press.

Plate-cutting:
The curved trusses at the Beijing National Theatre, China, designed by Paul Andreu, 2008, have been created by cutting plate steel. This has ensured that the many trusses are identical in size. Fully remediated welding has been used to connect the parallel chords and the web members. The horizontal connecting members were fabricated from round HSS that have been discreetly connected by solid steel half spheres.

Tokyo International Forum, Japan, designed by Rafael Viñoly, 1996. The fabrication methods were tuned to the capabilities of the period. The webs of the large curved beams were cut to shape and the flange plates rolled to fit. Many of the elements, though complex-looking, were identical and the overall layering and texture of the space provides a level of visual complexity.

Bending steel is a very hand-crafted process. The skill and experience of the roller-bending facility is critical to ensure the precision of the product. As the three-point bending equipment is only suited to making consistent curvature, uneven curvature requires a degree of approximation: the curve is divided into segments that each have a consistent curvature. Segments can then be welded together using seamless splice-welding to create a larger element of irregular curvature. This method is most often applied to tubular material. If irregular curvature is desired for wide-flange sections it may be easier to use digital files to cut the webs to the desired curvatures and use either the self-weight of the plate steel for the flanges to relax into the desired curvature, or plate-rolling to bend the flange steel. It is not uncommon to see a variety of such methods applied to a project, chosen as a function of the member types and the various distances to view.

2. DEGREES AND VARIATIONS IN CURVATURE

An important aspect of rolling and bending processes is the limitation on bending as a combined function of the degree of curvature and the thickness of the material. For non-symmetrical section types like rectangular HSS, channels and wide-flange (Universal) sections, orientation is also critical. This is often referred to as "hard way" or "easy way" bending. Bending in a direction that is more difficult may more likely lead to deformations if the chosen radius is too tight.

For structural sections, in particular pipes with a large diameter and thick walls, it may be necessary to use an industrial process called induction bending that uses heat to soften the steel. Here as well, if the bending radius is too tight, materials with thinner walls may deform, which is not good for aesthetics. Also if sections must be spliced, irregularities resulting from the bending process can result in deviations in the tolerances, meaning that the adjacent surfaces may not align.

While standard bending methods for structural sections and plates are most suitable to even curvature, parametric design promotes variations in curvature. To convey the impression of varied curvature it is possible to break the entire curve into arc segments, give each segment a relatively uniform curvature and splice segments via fully remediated butt-welded connections. This type of approximation requires that the splices occur at points where the curve is either not excessively tight or where the curves share the same linear tangent point in order to achieve more visually seamless transitions.

BROOKFIELD PLACE PAVILION

NEW YORK CITY, NY, USA, 2013

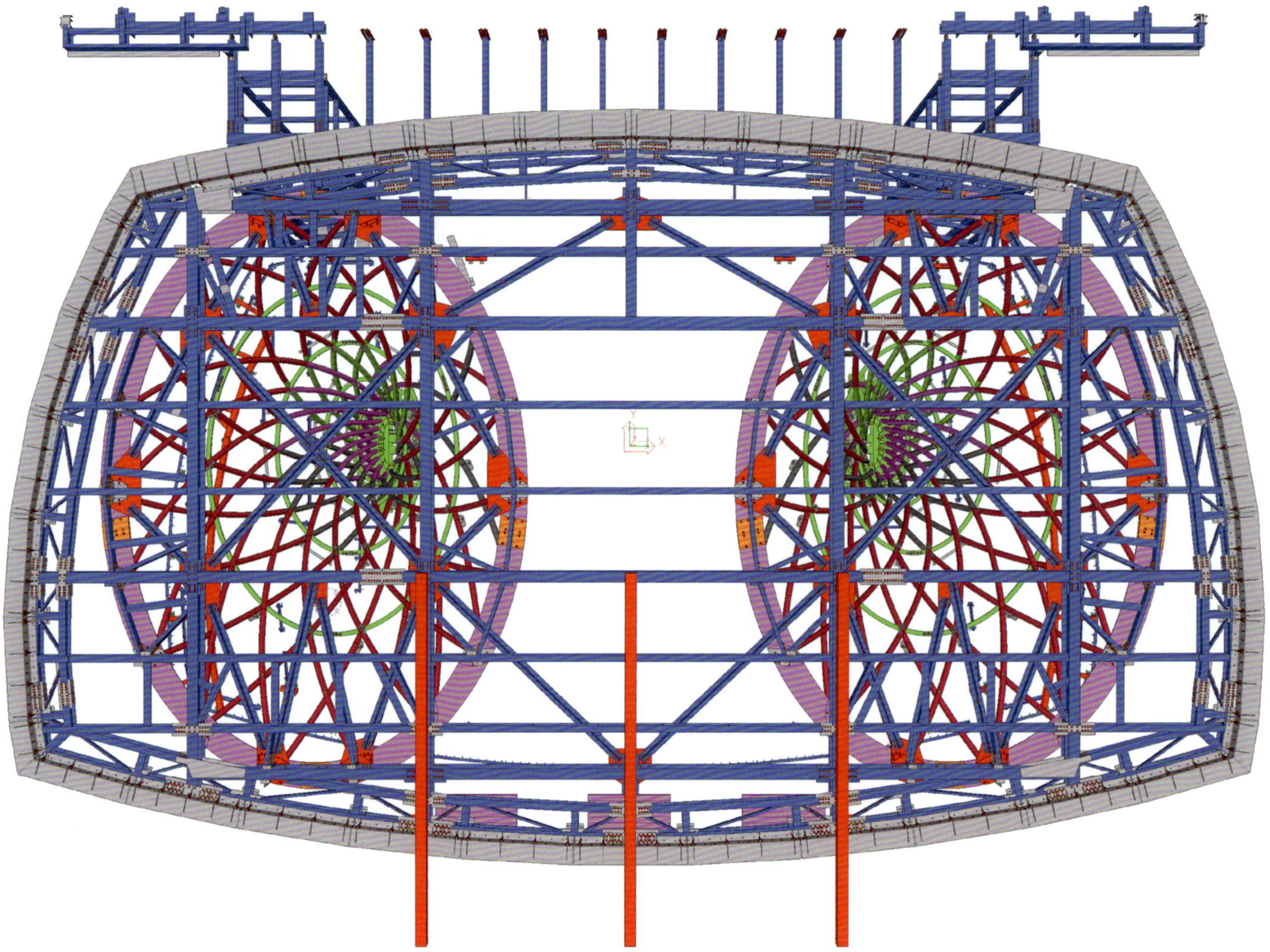

This plan view of the Tekla Structures model of the diagrid baskets illustrates the challenges due to the use of elliptical forms (which are more challenging to plot out at large scale), the varying nature of the curved HSS tubes as well as the asymmetrical nature of the support points at the bases of the baskets. The repetition of curvatures is minimal, as the symmetry occurs between the baskets, not within the individual basket.

ARCHITECTS:
PELLI CLARKE PELLI

ENGINEERS:
THORNTON
TOMASETTI

STEEL FABRICATION/
ERECTION:
WALTERS GROUP/
METROPOLITAN
WALTERS

Aspects of this project are addressed in *Chapter 4: Fabricating the Steel* and *Chapter 6: Erection Logistics*. There were very particular challenges in this project, owing to the nature of curved steel to be incorporated. A combination of bending methods was used to form the tubular steel of the diagrid baskets. The elliptical rings that constrain the multiple levels of the five-storey baskets were created by plate-cutting.

The baskets required careful tracking of the bent tubular members, as the geometries vary greatly around the elliptical curved elements. Here, the curvature is being verified by hand to see if it meets the digital specification. The tubes must be cut to meet the transportation requirements.

The elliptical rings that define and constrain the shape of the diagrid baskets were CNC-cut from plate steel. Circular indents allow the bent steel tubes to nest for welding. Temporary shoring in the shop is required to correctly position the elements in 3D space in preparation for welding. It is challenging to properly determine the positioning in the absence of any pure vertical reference member within the element.

It is important that the curved lengths of steel are properly tracked so that splices can occur within members that are of the same bending. There is some loss of material due to the cutting processes and preparation for welding, but this should be imperceptible. Here the welded splices have been distanced from the welded connection to the elliptical plate.

When fabricating for complex curvature a pre-fit test is critical to avoid issues on the site. The baskets were fabricated from five tiers of woven grids. While the base could be shipped as a unit, tier 2 was divided into two primary pieces, tier 3 into four primary pieces, and so on. Due to the overlapping weave of the tubes, X-shaped elements composed of two tubes were required to be installed between the major segments.

THE SAGE GATESHEAD
GATESHEAD, UK, 2004

The Sage Gateshead is a theater complex in northern England. The design concept involved the creation of a large, curved, shed-type roof enclosure comprised of Architecturally Exposed Structural Steel, opaque insulated metal panels and glass. The theaters within were constructed as independent reinforced concrete volumes that sit entirely within the shed enclosure.

ARCHITECTS:
FOSTER + PARTNERS

ENGINEERS:
MOTT MACDONALD

STEEL BENDING:
ANGLE RING METAL BENDING

FAÇADE:
WAAGNER BIRO

Resolving the Complexity of Double Curvature

The large shed roof that forms the basis of the design of this project is curved in apparent non-uniformity in its longitudinal and transverse directions. This means that it cannot use simple repetition of its curved elements as in a traditional barrel vault form. The majority of the curved members are unique.

Steel structural systems are quite naturally comprised of layers of structural elements, from those that are spaced closely to provide frequent support points for thinner cladding elements, to the very large members. Normally columns or frames provide the major load transfer paths to the foundations. In the case of the Sage Gateshead, the primary structural support system consists of four very large curved arches spaced along the length of the building. There is a secondary perpendicular system of smaller, curved, wide-flange beams spaced more closely to support another, even more tightly spaced system that provides the connective support to the cladding. Additional tension braces provide X-bracing between the large primary and smaller secondary curved beam systems.

Although the overall appearance of the building suggests a complex derivation of the geometry, the design-to-fabrication software available at the time of design and construction (1997 to 2004) was not advanced enough to support extreme irregularity. Complexity was achieved through the careful use of regular circular geometries that were laid out in a controlled plan system, which was subsequently rotated around a vertical axis to create the undulating volume.

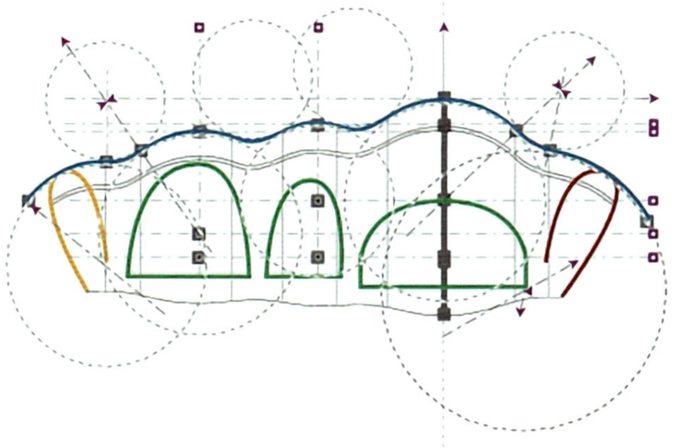

Although the overall exterior form of the Sage Theatre appears complex, it was derived from regular curved geometries that intersect at tangent points. The green lines denote the locations of the theatre volumes.

The layering of the structural system is clearly visible at the underside of the curved shed roof. Of benefit to the economy is the use of straight tubular members to join the smaller, curved wide-flange beams that connect the primary supports. These smaller round supports allow the curvature of the larger beams to dominate the appearance of the structure. The interior roof layer of galvanized steel decking is directly supported by this primary system.

The steel connections use an approach to AESS that highlights the connections and allows for bolting for on-site connections. This approach succeeds in accelerating the construction by decreasing the need for on-site welding.

The sizes of the members are tightly controlled. The dimension of the round base plate for the tubular layer fits neatly inside the flanges of the secondary, curved beam system. Simple bolted connections are used to join the steel. The curved wide-flange sections are fitted with end plates that have had welds ground in accordance with a higher-level AESS category. Again, bolting is used to join these sections, both for construction convenience as well as in acknowledgment of the distance to view, making the use of welding unnecessary. The glazing and cladding are separated from the main steel support system via a series of plate supports situated at the intersections of the round and wide-flange members.

Bending the Steel

This project used a hot-bending method more normally used for pipe members for large industrial or bridge applications. It was the best choice for the tighter curvature of the large wide-flange (Universal) members.

According to Angle Ring Metal Bending, the firm retained to bend the steel, two beam sizes of high-grade steel S355J2G3 were used for the primary supports, one being 838x292mm/33x11.5in and weighing 226 kg/m (151.8lb/ft), the second measuring 762x267mm/30x10.5in and 197kg/m (132.3lb/ft). Lengths of 18.5m/60.7ft were bent using hot- and cold-bending technologies, as suited the section types and radii.

Façade Design and Installation

The design of curved façades is addressed either by providing a genuinely curved surface or by approximating the curvature through the faceting of planar elements. In the case of the Sage Gateshead, façade specialist Waagner Biro employed both methods. A large portion of the curved façade is comprised of opaque stainless steel insulated panels, curved to fit the curvature of the secondary steel framework. The glass panels, by contrast, were conceived as planar, which greatly reduced their cost. In the early 2000s, when this project was completed, curved insulating glass panels were quite uncommon.

While the opaque insulated metal panels were formed with a slight curvature, the glazing panels are planar. Many of the opaque panels are located in the tighter radius of the valleys of the shed roof.

The large wide-flange steel members of the vertical support system have been formed using a hot-bending process. The glass panels have been fabricated as flat insulated sections and adapted to curvature by a secondary system of arms that attach back to another layer of steel that is curved along the length of the building and connected vertically by straight tubular steel members. The connections between the horizontal, curved wide-flange members are placed at the midpoint of the tubular system in order to allow for the bolting of the round end plates of the tubular members to the web of the beam.

SOUTHERN CROSS STATION
MELBOURNE, AUSTRALIA, 2007

The construction commenced with the erection of the Y-shaped columns, followed by the undulating triangular trusses that span between. Temporary vertical supports can be seen under the truss (at the left), as it was not stable until the joining roof elements were in place. Butt-welded splices were used to join the segments of the trusses. The curved HSS tubes that span between the trusses were erected one by one, then connected to each other with smaller tubular elements to form the roof frame. All of the connections of the roof elements used discreet bolted connections for erection efficiency. Australia has a preference for bolting at height to ensure worker safety.

ARCHITECTS:
GRIMSHAW
ARCHITECTS

ENGINEERS:
WINWARD
STRUCTURES

This railway station is dominated by an expansive roof designed to look like a stretch of dunes, intended to create a visual connection between the city center and the nearby Docklands. The station comprises an entire, trapezoidal-shaped city block, adding geometric issues to the planning. The undulating form of the envelope was developed in response to the hot external climate and the internal need for extraction of diesel fumes as well as to encourage ambient cooling via natural ventilation. The sides of the facility are open-air, with smaller conditioned facilities housed below.

Although the curved geometries are irregular in form, there is rhythm and repetition: of the Y-shaped column supports, the general geometry and curvature of the triangular trusses and their connecting tubular elements. The wave-like form is repeated for most of the roof, with common connection types throughout.

Discreet bolted connections were used between the irregularly curved HSS tubes that span between the triangular trusses. The color palette is muted and works well in a transit facility to lower maintenance costs by masking natural soiling. The underside of the roof is a lighter color, which assists in reflecting light back into the train bays.

Discreet bolted connections have been used to join the round HSS roof elements to the fully welded triangular trusses, also fabricated from round HSS tubes. The fillet welds have been left unremediated, given the viewing distances and multiplication factors. Complex cutting was required for the joining of the tubes.

The large hollow steel columns have been brake-formed to create their taper. The Y-shaped arms that reach out to support the sides of the trusses have been fabricated from plate steel and have fully remediated face welds. Plate sections have been joined as well as fully remediated crisp corners. Three-point bending has been used for the bottom chords of the triangular trusses that span between the columns.

Faint traces of the brake press can be seen translating through to the surface of the tall tapered steel columns. The columns consist of multiple sections of curved plate steel, due to the limit of width that could be accommodated in the brake press. The horizontal butt welds were fully remediated.

GARDENS BY THE BAY SUPERTREES
SINGAPORE, 2012

A main feature in Gardens by the Bay is a series of 18 "supertrees" ranging in height from 25 to 50m/82 to 164ft, distributed around the gardens to provide shade in this extremely hot climate. Several of the larger trees may be ascended and are connected by a complex suspended walkway.

The geometry of the structures is intentionally erratic to simulate a tree-like composition. In order to keep the cantilevered portions of the branches as light as possible, yet stable, lighter tubular steel rings have been added that tie the branches together in a very regular fashion. Color has been used to make the evidence of the rings recede – the branches a fuchsia tone and the rings painted white. Eventually, when the fast-growing tropical plants consume the structure, this differentiation will be unimportant, but in the initial years it is important to keep clarity in the structure. The rings also serve to stabilize the upper part of the structure both during the erection process and in the permanent condition.

A specialized lift system was installed inside the hollow concrete trunks of the larger "trees." The canopy elements were assembled on grade around the trunks and the entire top part of a tree lifted at once, the lift taking approximately 3.5 hours. Steel rings comprised of plate around the trunk kept the canopy element in alignment and were later used to permanently affix the steel construction to the concrete trunk.

Connection types are particularly varied in this project, as a function of the innovative intentions of the various parts of the structure. It makes clear use of layering for achieving higher-level stability while down-playing visual complexity. This is supported by the use of vivid colors to define the structural layer types.

ARCHITECTS:
WILKINSONEYRE,
GRANT ASSOCIATES

ENGINEERS:
ARUP

Each of the "supertrees" is comprised of a hollow concrete trunk surrounded by tree-branch elements made of round HSS tubes. The form bells outward at the top to follow the form of a tree canopy.

Color is used again to make the details of the support structure for the suspended walkway less apparent. The bright yellow paint on the railing system sets the walkway apart from the "supertrees." The stainless steel cables create a spiderweb effect.

A view up the gap between the tubular steel structure and the concrete trunk. The plate rings used to facilitate the lift of the top are visible, as well as the square HSS tube structure that holds the fuchsia-colored tree elements away from the concrete to provide access around the base as well as space for the plants to grow.

There is structural clarity in the placement of the steel. The regularly placed, curved vertical elements are overlaid with a slightly more irregular set of diagonals, which serve to brace the system as well as lend a more organic feel to the steel. The white tubular rings keep the cantilevering form of the "branches" intact. At the top these have been replaced by cables to provide stability and at the same time a lighter visual appearance.

THE CHRYSALIS
COLUMBIA, MD, USA, 2019

The practice of Marc Fornes/TheVeryMany is focused on explorations in formfinding. They worked very closely with Arup to ensure that the unusual shape of the pavilion would not result in detrimental wind or snow loading. The structural frame was reinforced accordingly. The shingled aluminum cladding, fabricated by Zahner, consists of 7,700 pleated forms.

ARCHITECTS:
MARC FORNES/
THEVERYMANY
WITH
LIVING DESIGN LAB

ENGINEERS:
ARUP

STEEL FABRICATION/
ERECTION:
WALTERS GROUP/
METROPOLITAN
WALTERS

The Chrysalis was designed as a multifunction pavilion whose form could accommodate a wide range of public activities, both formal and informal, from high-end theater and music performances to more casual gatherings. The form was created around a series of arched openings, each intended to support a specific function, from a theater set to a loading bay.

Marc Fornes describes the design as follows: "To achieve a light and organic effect that suits the context of a dense wooded park, the studio took a structurally-oriented approach, building upon over a decade of research and development of lightweight structural shells that unify form, support and experience into a cohesive system. In particular, The Chrysalis further develops principles explored in its 'little brother' precedent, Pleated Inflation, completed by the studio in Argelès, France, in 2015. The Chrysalis is similarly generated from a process in which a digital mesh is drawn flat, and all of its segments are transformed into a series of differentiated spring systems, then inflated. Constraints for pleating are added to the inflation protocol to provide extra structural depth. Layered within the pleated shell, an exoskeleton was designed to support the heavy loads for performances inside, such as lights and other rigging. Engineered into steel tubing by Arup, it supports 70 point loads that can each sustain 1 ton."

The fabricator used Tekla Structures to detail the tubular frame. The superstructure of 10in/254mm and 8in/203mm tubular steel is knit together with moment connections, making the whole arched shape act as a singular structural element. 300ft/91m of 7in/178mm tubular steel was used to brace in-between the compound curved arches to be able to suspend up to 21 tonnes of performance equipment.

The custom bent steel segments were delivered to the site unassembled, meaning that considerable bolted on-site assembly was required. Temporary shoring supports were needed, as the small element size was unable to be self-supporting until erection was complete. Alignment and precision were critical to support the exacting requirements of the aluminum shingle cladding.

The compound curvature was created by using smaller, more uniformly curved segments of tubular steel that were spliced together using very standard plate-to-plate connections. The intention of the AESS was for a very technical and rugged aesthetic. This approach contributes to a high degree of flexibility for supporting the equipment required for a wide range of performance types.

The interior view of the completed pavilion shows the exposure of the tubular frame. The tubular arches underscore the overall form-driven aesthetic. The density of the smaller connecting members that provide rigidity lend a fabric-like visual complexity to the structure.

PHOENIX INTERNATIONAL MEDIA CENTER

BEIJING, CHINA, 2014

In a period when many international firms have been awarded projects of significance in China, the Phoenix International Media Center stands out as designed by a local Beijing firm. The contours were designed using parametric 3D modeling software, including Frank Gehry's Digital Project modeling tools, which enabled architect Shao Weiping to intricately manipulate the parameters of the overall design and structural engineering of the Phoenix Center. The intention was to control the airflow around the building and also convert the façade's steel lattice grid into a network of tiny canals that transport rain into an array of reflective pools that surround the building. The structure features a parametrically designed atrium, the entire project based on the concept of a Möbius strip.

The steel lattice of the highly transparent façade is a fully welded structure. All connections between the tubular sections have fully remediated welds to support the look of continuity upon which the idea of the Möbius strip is based. The interior steel is all AESS, while the steel that supports the façade is concealed by interior and exterior finishes.

ARCHITECTS AND ENGINEERS: BIAD UFO

The lattice structure of the Phoenix International Media Center is fabricated from a parametrically designed overlap of an interior layer of round HSS tubes and an exterior layering of custom-fabricated steel elements, which provide structural support and discreetly house the HVAC and electrical service runs.

This interior view during construction shows the fabrication of the ramp platforms from sheet steel that has been bent using brake-forming. The welded connections on the interior tubular system are visible. Temporary bolt tabs were provided, which were subsequently removed and remediated to create a seamless AESS finish. The transverse exterior lattice layer is also visible under construction. It consists of a ladder of welded tubular steel that has a welded box section connected towards the exterior. Space is included here for mechanical systems and enough depth to connect the elaborate façade system.

As the lattice structure was fully welded, platforms were required at each connection point. The inwards curvature precluded the use of exterior scaffolding. The part of the building shown here belongs to one of the two interior office blocks, which were helpful in the provision of lateral stability to the lattice frame.

The spiraling walkway is connected to the interior round HSS system via a series of round HSS struts, using pin connections on either end to allow for site adjustments.

The form results in significant complexity in the design of the cladding system. The 3,800 glass panels are each differently sized and detailed. The shingled look of the form is part of the control of rainwater and ventilation.

3. COMPLEX CURVATURE

Irregular curvature, large section sizes and a preference for members that are rectangular in cross-section make simple three-pressure-point bending unsuitable for many projects. In these instances plate steel is employed to create the components, using a combination of roll- and brake-forming. Digital studies determine the precise nature of the desired curvature of the element and whether the curvature is tight to a degree that requires plate-rolling or if the self-weight of the steel can be used to let gravity ease the plate into position. If the intended curvature of the plate is irregular, brake-forming is typically required. Where surfaces exhibit complex curvature, accurate digital sizing information is translated to cut the plate materials to the proper shape.

For members of rectangular cross section, the method often depends on the corner details – whether a sharp or rounded appearance is desired.

The curved shape of the Peace Bridge, Calgary, AB, Canada, designed by Santiago Calatrava, 2010, was fabricated from brake-formed plate steel – from the diamond formed ribs to the curved supporting ends. Overall there is a slight camber in this low-profile pedestrian bridge, designed to be a level crossing for cyclists and pedestrians.

The rectangular sections were tightly curved in multiple directions, meaning that plate-rolling alone would not have produced the correct curvature. The corners were intended to be fully remediated and slightly softened rather than sharp. Brake-forming was required. There are faint traces of the internal plate reinforcement showing (see Shenzhen Sports Complex on p. 114 for this detail).

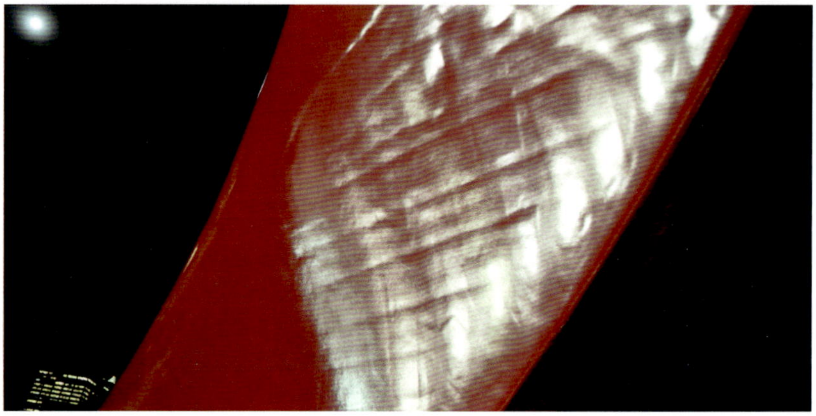

This night view of the steel reveals the slight translation of the hits of the brake press on the steel. This surface imperfection is not visible during the day but is a reminder that when using brake-forming great care needs to be taken in applying pressure to the plate material.

The curvature of the ribs of the Puente de Luz pedestrian bridge in Toronto, ON, Canada, designed by Francisco Gazitúa, 2011, is twisted. Sharp corners were desired. The curves were created via plate-rolling. An inset of the meeting faces allowed for fillet-welding, providing a nice shadow line and eliminating the need for excessive weld remediation at the corners.

The curved nature of the ribs that form the bridge required plate-rolling at the bending facility, with member fabrication at the fabrication shop. Digital models were essential to determine the plate sizes prior to rolling.

Although a sharp corner detail was desired on the Arganzuela Pedestrian Bridge in Madrid, Spain, designed by Dominique Perrault Architecture, 2010, it was felt that a rougher aesthetic was acceptable and that no weld remediation would be carried out. There was significant on-site welding required for the bridge, so the multiplication factor was an issue. The plates were allowed to meet sharply at the corners without any inooto.

The exterior and interior faces of the square member sections were rolled, while the sides were cut to curvature and remain flat. The dappled light coming through the many screened elements of the bridge complements the design approach to achieve texture.

The Shenzhen Sports Complex, China, designed by AXS Satow, 2011, is flanked on either side by a partially covered walkway that consists of a curved diamond grid fabricated completely from custom rolled steel. The design sought sharp edges to the meeting points of the plates that required significant precision and weld remediation.

Unfortunately, the project is suffering severe degradation due to insufficiency of its coating system. Show-through from the welding of the internal support system reveals the locations of the plates that provide additional rigidity to the face plates. Here, it becomes obvious that on-site welding was used to attach three of the four arms to the larger prefabricated element, so that the more difficult elements could be shop-fabricated. The curved faces would be plate-rolled.

Large expanses of the curved custom-formed lattice grid are supported by clusters of round HSS tubes that branch out to provide more frequent support for the lattice yet with minor interruptions to the flow of public space. The stiffness of the gently curved lattice is able to support a significant cantilever.

The fabrication methods required for the Oculus, New York City, NY, USA, designed by Santiago Calatrava, 2018, were very complicated given the irregularity of the structure, the height and the number of cantilevered elements. Extensive use of plate steel was required to fabricate the large external fin structures. The C-shaped elements on the faces of the fins were used to attach special clamps to secure the subsequent extensions in place for welding support.

The base structure for the fins consists of round tubes that form the edges of the fins, joined by welded plates on each side. An internal support system maintains the distance between the face plates under loading. The connection location has been stepped so that the incoming element naturally slots and is held accurately in position. Some weld remediation was carried out at the lower levels, but the face welds on the fins were not remediated.

The base fins were provided with temporary steel that provided points of attachment for custom clamps that were essential to secure the prefinished extensions in place during the extensive on-site welding process. Access for the ironworkers to this level could be provided by a lift truck. The overall work heights exceeded standard scaffolding so other innovative methods were required.

The horizontal tubular connecting elements that join the long fins can be either reached via a lift truck or were provided with custom working platforms for secure welding access. All of the steel above the simply primed lower levels arrived at the site with a partial white finish. Finishes were held back from the weld locations. This reduced the efforts required to apply the final coatings on the exterior elements.

ANGULAR GEOMETRIES

CHAPTER 9

The most direct result of the impact in advances in digital technologies that followed the shift in design towards Deconstructivism and then Parametricism has been the widespread incorporation of angular geometries in steel structures. Whether architecturally exposed or concealed, they define the overall building shape. Their implementation has had a great impact on the fabrication and erection process, as the steel industry has been able to respond to this development due to the interoperability of software, directly to inform fabrication processes and also to assist with erection sequencing logistics.

1. DECONSTRUCTIVIST BEGINNINGS

Unlike the projects addressed in *Chapter 7: Economically Driven Strategies*, many complex steel projects that use angular geometries push them to the forefront of the aesthetic. The fabrication of steel has been able to match the more geometrically challenging expectations of Deconstructivist and parametrically influenced designs. These form-driven designs are distinct from diagrid structures, which are based on a recognized and systematic design process that entails the use of nodes as connecting mechanisms. This will be addressed in *Chapter 10: Nodes;* diagrids are investigated in great detail in *Diagrid Structures: Systems, Connections, Details*. The angular geometries investigated in this chapter consist of less systems-driven and more chaotic or varying assemblies of angular steel.

Highly irregular angular geometries require extreme precision in their fabrication, with maximum work done in the shop, where purpose-built jigs can ensure that the steel exactly matches the digital model. Early angular projects had to rely on hand measurements to translate dimensions to the steel. CNC processes are essential to creating accuracy and efficiency in the translation of digital dimensions to the cutting of the steel. Tolerances must be held extremely tight, in particular for AESS projects with preferences for an all-welded look, as dimensional mismatches are difficult to conceal. Extremely large structures do not have the benefit of pre-fit testing and will be forced to resolve fit issues on site. Some workaround strategies include positioning potentially problematic on-site-welded connections so that view to them is distanced or obscured.

The Kurilpa Bridge, Brisbane, Australia, designed by Cox Rayner Architects with Arup, 2009. This tensegrity-based structural system was the first spanning structure of its kind. There is an underlying order behind the apparent and intended chaos of the structure. The member sizes clearly indicate their compressive or tensile designations. For more on this project see p. 128.

Steel fabrication software was in its early development for the design and construction of the Seattle Central Library, WA, USA, designed by OMA, 2004. The detailing of the meeting points of the many angular façade sections strove to remove these connection points from close view, as a means to make any imprecision less apparent. The irregularity of the faces of the library resulted in numerous asymmetric edge conditions.

The limitations of the interconnectedness of design and fabrication software in 2002 – 2004 are evidenced by the roughness of this corner detail, where a number of angular diamond-grid planes meet. From an architectural perspective it is reasonably well hidden from view. It is an aspect of steel detailing that can be better solved with current software and developed fabrication expertise.

2. CONCEALED SYSTEMS

If the ultimate intent is to conceal the steel, member and system choices are greatly broadened. Tolerances still need to be kept tight to ensure proper fit, but not in the same way as for AESS structures. The erector has options to adjust or push members into alignment without fear of damaging the surfaces or edges. More usual spanning methods, such as trusses, can be modified to suit the geometric requirements without aesthetic concerns. Cladding and interior finish systems may employ intermediate support systems to build out their final profile to more accurately fit to the digital model, allowing a more lenient fit to the structure itself, as discussed in *Chapter 7: Faking the Curve.*

This interior view of the Dalian International Conference Center, Dalian, China, designed by Coop Himmelb(l)au, 2012, shows that the undulating, parametrically derived steel structure is completely clad, thereby drastically reducing the requirements on the steel detailing. A triangulated system of cladding also removes the necessity for true curvature from the project.

A variety of approaches was taken to achieve the erratic steel forms, as evidenced by this construction photo. Although on-site welding is a common practice in China, bolting could also have been used, given the concealed nature of the structure.

3. MODIFIED GEODESIC SYSTEMS

Triangular geometries are often used to support curved forms that are not necessarily spherical in nature. These function as a variation on the geodesic dome, which is a hemispherical thin-shell structure whose triangular elements are structurally rigid and distribute the stresses throughout the structure, making them able to withstand very heavy loads. They also function as a variation of genuine diagrids, whose triangulated system carries the vertical and lateral loads, with horizontal tension rings to resist bursting forces. The triangulated systems referred to here have a much more uniform selection of member type, and the scale of the triangulation mediates between the approximated curved form and the desired curvature of the cladding systems on the interior and exterior. The planar nature of these modified geodesic systems somewhat reduces fabrication and erection complexity and challenges, as alignment and fit are potentially more predictable.

The Museum of the Future in Dubai, UAE, designed by Killa Design, under construction in 2018. A round HSS frame negotiates the overall curved shape of the building. The intention is to conceal the steel on the exterior and interior behind truly curved systems. Scaffolding fills the central void during construction to provide access for welding work.

The majority of the members are straight. Occasionally, where the curved form is more constricted, a bent segment is used or, as pictured here, round sections are cut and welded to include an angle. All-welded connections were selected in order to keep connections very tight to ensure minimal interference with the cladding systems. An extensive scaffold was required to provide access for the welding and finishing processes. Temporary bolting tabs can be seen. Minimal remediation will be needed, as the system is concealed from view.

The general increase in the section size of the members allows for significantly increased loading compared to earlier geodesic systems, whose original intention was to support a relatively lightweight enclosure system. The modified geodesic structures are capable of supporting interstitial floors and other more typical building loads, in addition to cladding.

The AAMI Park Stadium in Melbourne, Australia, designed by Cox Architecture with Arup, 2010, uses a modified geodesic dome to create covered seating around the perimeter. The large scale of the triangulation is in keeping with the large section diameter of the HSS members. The large size of the triangulated panels creates a more faceted appearance.

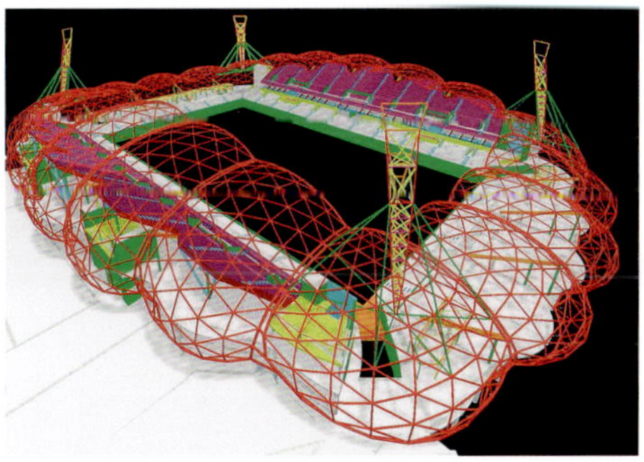

This Tekla Structures model shows the overall modular geometry of the modified geodesic system. The triangulated method using customized components allows for modulation of the forms to create valleys between the rounded sections. The increased articulation of the forms also adds strength, as the shells cantilever out to protect the patrons from weather and column supports would interfere with sightlines.

Discreet bolted connections have been used for the round HSS frame on the interior of the canopy between the larger assemblages. The cladding fits tightly to the frame via a series of clips bolted to plates that have been welded to the HSS frame.

4. ARCHITECTURALLY EXPOSED ANGULAR GEOMETRIES

Exposing the steel adds greatly to the challenges of fabrication and erection with highly irregular angular geometries. When the structure ceases to be planar and depth is also added, accurate translation from the digital to the constructed is essential. The fit issues will be more consequential when planning for irregular angles. Whereas modified geodesic forms have a high degree of modularity to decrease unique instances, chaotic geometries often have many unique elements and assemblies.

FEDERATION SQUARE
MELBOURNE, AUSTRALIA, 2002

Federation Square was designed at a time of transition from theoretical to constructed deconstructivist architecture and looked to reflect the influence of Deleuze's anexact geometries. Stan Allen wrote of the project in 1999: "This is to say (as the Assemblage project draws to a close) that the most urgent and interesting issues are no longer debates about formal language, its origins or affiliations, but rather, the complex questions attending its realization in built form. Debates about blobs versus boxes dissolve in the common filter of technical information."[6] The challenge was to find ways to transform the concept for the chaotic angular steel frame, which had been modeled in wood for the competition submission, into a constructable steel frame. In a marked change of working method, the firm invited the engineers and consultants to work with them in their offices.

The structure consists of approximately 2.1km/1.3mi of square HSS sections, weighing in at 3,880 tonnes. Galvanizing was chosen as a corrosion system because the frame is extensively exposed to the weather, but also because the nature of the finish complements the rugged connection detailing. Much thought was given to transforming a connection detailing initially conceived in a glued wood-frame model to maximize shop fabrication. The 3D nature of the structural system assisted in the creation of self-supporting elements that required little to no temporary support during erection. Innovative discreet bolted connections were used throughout to also facilitate more efficient erection.

ARCHITECTS: LAB ARCHITECTURE STUDIO WITH BATES SMART

The overall depth of the 3D structural frame in conjunction with the triangular pinwheel pattern of the cladding system achieves an even higher level of visual complexity than either system could achieve on its own.

[6] Bates, Donald L. and Peter Davidson. "Federation Square, Melbourne, Australia." *Assemblage* 40, 1999.

For the support of the angular planes of the exposed galvanized steel decking, a simpler orthogonal system of beams and purlins is used. A very light system of X-bracing is visible, which adds lateral support to the roof.

At the points where the planar support system is attached to the irregular support of the 3D structural frame, a system of round HSS struts of varying lengths is used. This allows for a great deal of flexibility in the interface between the uneven geometry of the structural frame and the cladding system and also is able to use more standard repeated details.

The intended three-dimensionality is achieved by a combination of the steel frame with a series of more standard base elements. The Y-shaped portion of the frame is sitting in behind and was shop-fabricated as a flat component. The angled members attached in front have a special fitting that allows for angle cuts in the plates to more easily accommodate variation in angles of attachment. This method allowed some uniformity within a chaotic spatial assembly. Fillet welds have been used at these connections. Discreet bolted connections were used for on-site connections throughout the project.

BMW WELT MUSEUM
MUNICH, GERMANY, 2007

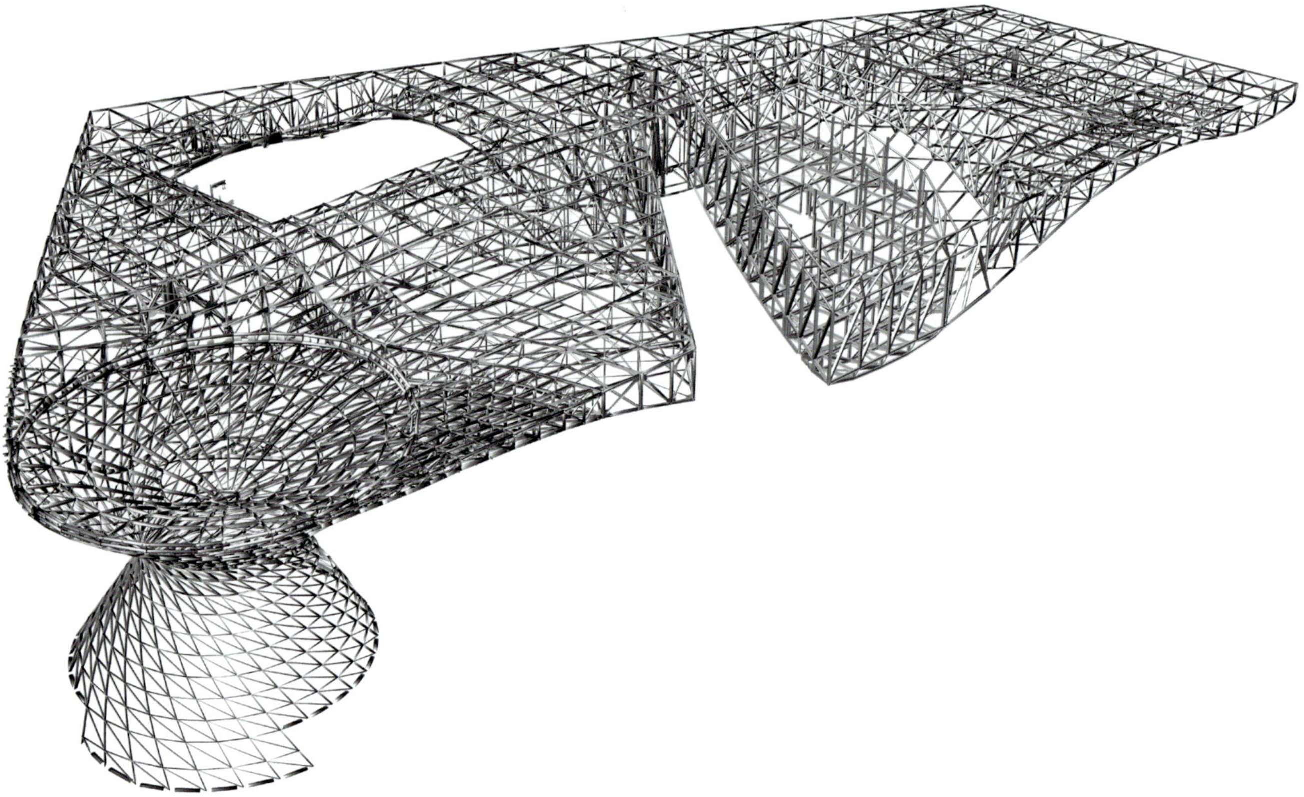

This 3D view of the steel structure of the museum shows a clear differentiation of the nature of the AESS steel used for the double cone and the two-way truss system of the cloud, which comprises the undulating roof of the main building. The steel for the primary roof is concealed and was designed to achieve long clear spans in order to minimize column interruptions to the flow of the space, in addition to supporting a large hanging walkway.

ARCHITECTS:
COOP HIMMELB(L)AU

ENGINEERS:
B+G INGENIEURE, BOLLINGER UND GROHMANN GMBH

"Today, everything is conceivable, and depicting daring architecture in colourful announcements is easy. But what is still radical is wresting these images from a one-dimensional illustration and pushing them through – and realizing them – in three dimensions. Radical architecture is only radical today when it is also built… that is the difference between now and then." Wolf D. Prix

The design and fabrication of the BMW Welt Museum was able to take advantage of significant advances in the software used in steel design. Nonetheless, the project is clearly divided into two distinct zones. A double cone geometry has been used as a major visual support to the 16,000m²/172,222sf steel truss "cloud" that is intended to appear to float on a structure comprised of only 12 columns. The cloud consists of an upper and lower grid layer of 5m by 5m/1.5ft by 1.5ft cells. Diagonal struts between these layers interlock the grids into a spatial supporting structure. The double cone is an AESS feature that used HSS sections to create its tornado-like twisting pattern. The "cloud" is completely concealed so that wide-flange sections with extensive use of bolted on-site connections could be used. Custom plate details accommodate the many odd points of intersection between the "cloud" and the double cone.

Photographer Marcus Buck spent an extensive amount of time documenting the construction of the museum. Here, we can see the double cone complete as the welded connections are also being completed. The structure was fully engulfed in scaffolding to facilitate access to the numerous welded connections.

The double cone houses the major exhibition space and forms a major support point for the roof.
The construction of BMW Welt required 4,000 tonnes of steel. About a quarter of this was built into the double cone alone. With a height of 28m/92ft and 48m/157ft in circumference, it reduces to 14m/46ft exactly at its center. Every single steel section in the double cone was manufactured with its own unique template. Tolerances were held within 2mm/0.08in of the design specifications. These sections also function as hidden ducts for data cables and so their interior continuity was critical. To facilitate the extensive on-site welds required, the double cone was completely consumed with scaffolding.

Buck's views are indispensable to provide a better understanding of the realities of the erection process. A scaffold tower supports the conical roof of the double cone and provides access for ironworkers to complete the connections. As the steel here will be concealed, bolted wide-flange (Universal) sections have been used. The HSS sections that comprise the double cone structure are visible below. The rust-colored primed steel visible between the grey primed locations denotes the locations of the fully remediated welded connections.

CANADIAN MUSEUM FOR HUMAN RIGHTS

WINNIPEG, MB, CANADA, 2014

The form of the museum, its massing and material use created extremely distinct approaches to the detailing of the steel structure.

The complexity of forms and materials comprising the design of the Canadian Museum for Human Rights required an extremely high level of coordination between the architect, engineers and steel fabricators. The design for the building started as a sculpted clay model, meaning that the massing and its inspiration were extremely organic in nature. Revit and Tekla Structures were used by the team to create as seamless a workflow as possible, interpreting the conceptual ideas behind the project and transforming these into a constructable project.

The steel design took on very distinct forms in the various parts of the building. Generally the idea of human rights led to a structural interpretation that leaned towards rugged detailing. The sloped mass of the building, to be clad in stone, used a diagrid structure fabricated from wide-flange sections with large steel plates to act as the nodal connections. The truss-like framing that supports the large glazed portion of the building was designed as an AESS system, albeit with quite rugged detailing to reflect human suffering. The Tower of Hope that ascends through the middle of the building was fabricated from galvanized steel, as it is exposed to the elements at the top. Details of this element were included on p. 41.

ARCHITECTS:
ANTOINE PREDOCK
WITH SMITH CARTER

ENGINEERS:
HALCROW YOLLES

STEEL FABRICATION:
WALTERS GROUP

The partially completed steel frame during construction in 2011. The diagrid tower to the right was to receive a stone cladding and so have its detailing concealed. The more loosely spaced steel frame to the left would receive a glazed façade and therefore was detailed according to AESS requirements.

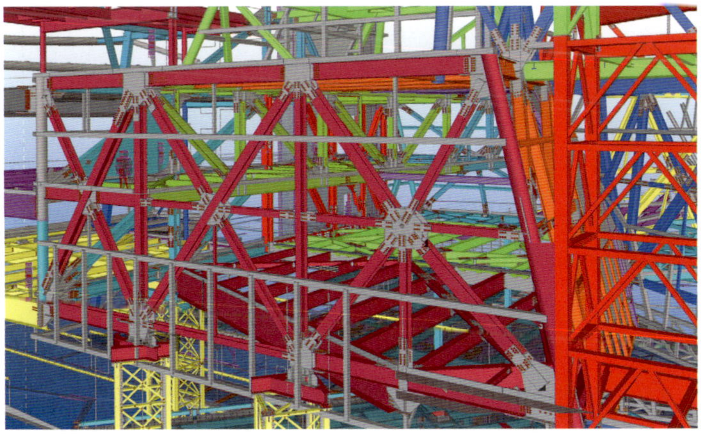

The Tekla Structures model is used to look not only at the detailed design of the many complex connections, but the coloration also indicates a plan for the erection sequence. These models are also used to coordinate with the mechanical and electrical services.

Where the steel is concealed behind the support system for the masonry wall, the use of large plate connectors to facilitate the erratic angles of the nodal connections allows for a compact transfer of forces. The inconsistency in the connections was readily solved by the software but also did not require an aesthetically driven approach to detailing.

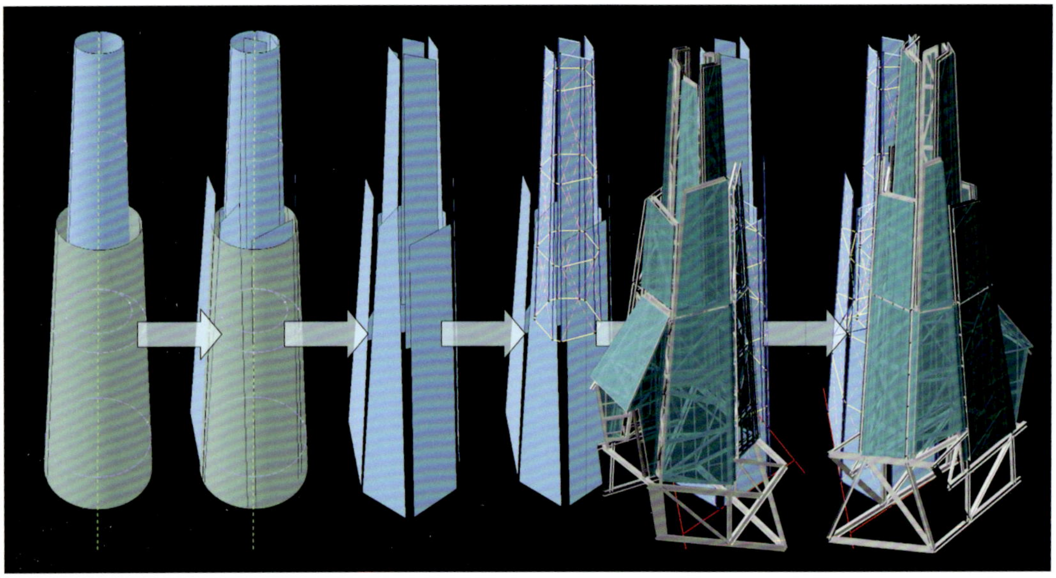

The challenge in the steel framework for the Tower of Hope was to start from more normalized geometries and evolve them into a series of planes and frames that could be resolved into a structural system.

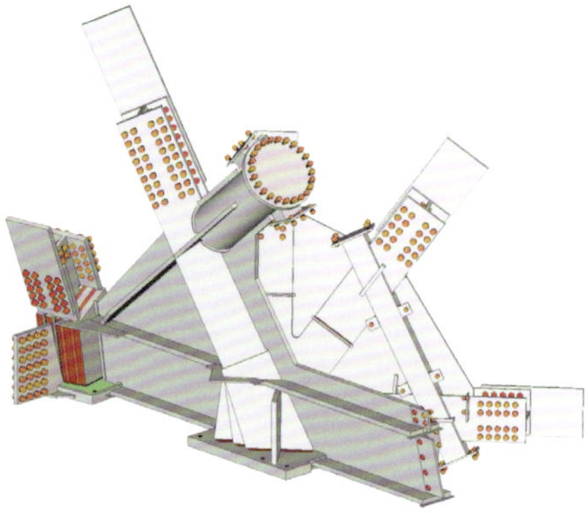

The ability to digitally model the complex steel nodal connections was critical to solving the complicated load path requirements and stress concentrations in this chaotic structure. Much of this innovative problem-solving is concealed from view in the final project.

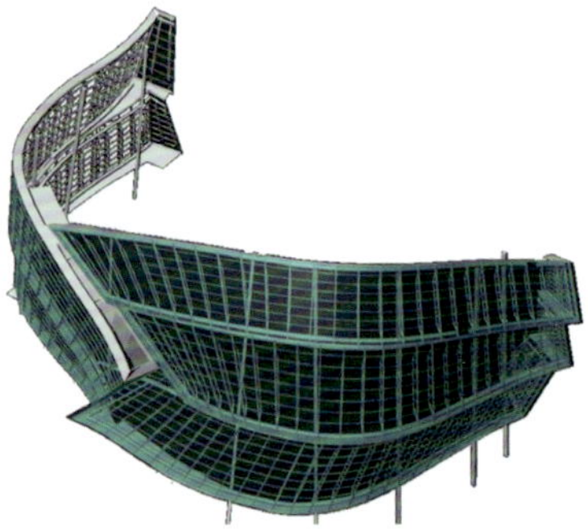

The curved glass façade is supported by round tubular elements referred to as "cloud rails," as the glazed portion of the building was referenced as a "cloud." The limits on the diameter of this tube were based on the maximum capacity for bending equipment, combined with the maximum standard fabrication diameter of HSS tubes, which ranges from 42 to 44in/ 1 to 1.1m. As the curvature is irregular and the rail extremely long, it consists of spliced elements.

The "cloud rail" is part of the AESS system that provides support at the base of each section of the large curved façade. The detailing of the HSS steel trusses that support the glass is far more refined, and the elements are regularly spaced to align with the glazing mullions.

Far more regular steel systems have been used to frame the opaque roof elements. A radial pattern of trusses fabricated from wide-flange (Universal) sections supports standard corrugated decking. The variation in the ruggedness of the detailing of all aspects of this structure fed into an aesthetically driven concept based on the issues of human rights.

The round tubular "cloud rail" supports a series of vertical trusses fabricated from fully welded round HSS members, which are aligned with the vertical mullions of the glass for direct wind support. A more angular vertical truss system, custom-fabricated from plate steel with the appearance of wide-flange sections, supports the "cloud rail" and the roof. Although there are bolted connections visible on these vertical supports, the detailing is somewhat suppressed compared to the more chaotic steel detailing that supports the core and tower of the building.

KURILPA BRIDGE
BRISBANE, AUSTRALIA, 2009

The erection started with the sections immediately over the concrete supports. The tension members were used to support the builds as they cantilevered out. The middle section was prefabricated and lifted into place to complete the bridge. The fast flowing river precluded the use of a temporary support system in the river. The spans are 58m, 125m and 45m/190ft, 410ft and 148ft.

ARCHITECTS:
COX RAYNER
ARCHITECTS

ENGINEERS:
ARUP

The Kurilpa Bridge is the first-ever bridge to be designed as a functioning tensegrity structure. The idea evolved out of a combination of several needs: to maintain a fairly level crossing of the deck to provide for accessibility, to keep the height profile lower than would be afforded by a more standard mast and cable system, to maintain the overall depth of the bridge deck to no more than 1m/39in for river traffic clearance, and to complement the nearby Gallery of Modern Art within the Millennium Arts Precinct.

The base configuration started conceptually as a more regular series of round vertical masts, slightly higher over the concrete supports and connected by cables. These were then raked and also joined by additional cables. Another layer of horizontal flying struts was tied to the raked masts, supported tensegrity-style, however with cables. The additive layering of the structure, with carefully coordinated offsets, was intended to create a visual randomness and chaos.

A continuous shade awning was a requirement for the crossing. This is suspended from pyramidal tubular HSS frames that are also hung from the mast system via cables.

The horizontal tubular cross members, termed flying struts or spars, are held in place only by cables. A series of pyramidal-shaped round HSS structures have been suspended from the primary raked masts, from which the continuous shade cover is hung. The raked masts are custom-fabricated tubular steel sections up to 30m/98.5ft long, with section sizes varying from 610mm/24in diameter to 905mm/36in diameter.

Discreet bolted connections have been used to connect the angular round tubular compression members to the primary spanning beams. Bolted connections have also been used to attach the deck components. The deck beams are rolled-steel I-sections, typically 530mm/21in deep. Where the clevises of the primary tension members are attached to the main support beam, donut-shaped steel elements have been added to the plate detail to increase the overall thickness, so as to respond to the pull-through forces rather than thickening the entire plate. This creates a visual enhancement to the detail.

The clevis fittings and cables have been fabricated from either galvanized or stainless steel to respond to corrosion issues. There is a variation in the diameter of the cable to suit the degree of tensile loading. Likewise, the plate thicknesses at the attaching points have been varied according to loading, such that the member selection and detailing is able to tell a story about the forces. The major cables are high-strength spiral-wound galvanized wire ropes 30-80mm/1.2-3.2in diameter.

A splice was necessary in the custom tubular compression members, as they were 30m/98.5ft long. While the longitudinal weld seam has not been remediated, the butt weld between the sections has been. Although this detail is well removed from near view, it was important for the overall visual lines of the supports to maintain the continuity of the linear seam appearance and avoid a shadow line from the tube-to-tube connection.

NODES

CHAPTER 10

The steel structures that defined the Modern era were characterized by fairly standard "framed" connection details. Standardized ways of connecting beams to columns were possible due to the orthogonal nature of the structure. Even the engineering of many structures was able to be simplified into a two-dimensional system of forces. The advent of complex structures introduced a high level of difficulty both in the calculation of force systems as well as the means to resolve their transfer through connections points. This was due to the more three-dimensional nature of the overall form as well as the absence of right angles of intersection. Complex geometries require a more responsive approach, and nodal connections have been developed to satisfy this role. Nodes play a critical role in the design of spaceframes, diagrid structures and lattice/gridshell systems.

1. THE EMERGENCE OF THE NODE: SPACEFRAMES

node /nōd/
Noun. A point at which lines or pathways intersect or branch; a central or connecting point.[7]

[7] Definition of Node by Oxford Dictionary on lexico.com, a common definition referring to structural systems.

Although in a looser sense the terms "joint" or "connection" might be considered synonymous, the term "node" conjures a very different visual idea of the aesthetics and functional requirements of a three-dimensional connection. Nodes as structural and functional terms first appeared in conjunction with geodesic domes and spaceframe technology. Nodes were critical in the creation of a heavily prefabricated system that could permit easy assembly on site. The members were fairly light and the geometries, though relatively regular, required the resolution of a high number of angled components at a point. The size of the node was directly related to the number of connecting members and the ability to nimbly accept their summative cross-sectional areas. Traditional steel fabrication methods using standard bolted or welding methods were incapable of addressing this type of connection. They were too bulky.

The incorporation of cast steel nodes at the intersection of the 8 concrete filled steel tubes that form the delta frame supports at the Queen Richmond Centre in Toronto, ON, Canada, 2015 by Sweeny&Co Architects was a critical aesthetically driven choice that has resulted in the numerous design awards that this project has received. The project required a high level of fabrication coordination between the casting designers CastConnex and the steel fabricator/erector Walters Group.

Spaceframes were designed as prefabricated systems subjected to fairly consistent loading that encouraged significant uniformity in their design. Although some pre assembly of larger spaceframe assemblies can be done prior to lifting, many of the connections are done on site. Spaceframe nodes were quite unlike the angled connections in 2D or 3D trusses which were normally shop-fabricated and could afford some structural continuity of the chord members through the joint, providing stiffness. In theory, moment resistance at panel points is not required, as truss members are typically designed only for axial loading. Nodes, on the other hand, tend (although not exclusively) to be fabricated as discrete elements to which other members connect, so there is a level of discontinuity at this point, shifting certain responsibility to the physical connection between the node and the incoming members.

The ball-type node in a spaceframe is able to accept a large number of incoming connections. The structural requirements of the spaceframe elements are accommodated by a variation in the member size. Special end connections facilitate the member-to-node connection.

Ferrari World in Abu Dhabi, UAE, 2010, by Benoy Architects is an example of the application of curvature to a spaceframe, possible due to the immense scale of the building. The structure remains highly regular and repetitive.

Spaceframe structures never succeeded in making a large impact in mainstream design, as there was resistance to the very technical and regular aesthetics of the type. In the same way that High Tech style was resisted until it evolved into Architecturally Exposed Structural Steel, spaceframes were not seen in design terms until their flexibility improved in formal terms. Although spaceframes need not be perfectly planar and can incorporate curvature if the building scale is large enough, most architects tended to avoid the systems. Spaceframes have found uses for large-span cases such as airports, convention centers and exhibition buildings. As referenced in *Chapter 7: Economically Driven Strategies*, spaceframes have recently been used in projects such as the Beijing Capital and Shenzhen International Airports, though partially obscured to "improve" their aesthetics. The digitally driven expression that is found in complex steel structures and the exploration of chaotic and irregularly curved geometries doesn't align well with highly regular prefabricated systems. These geometries demand a more responsive solution to connection design.

The geometry of the spaceframe created different demands on cladding systems, as standard aluminum-framed curtain wall systems were not easily adapted to the variation in geometries, particularly early on, prior to advances in digital modeling. Additionally, the spaceframe structure was looking for lightness, and traditional glazing systems were also very heavy and so not the best choice for the vision system. This offered the potential for developments in newer ETFE systems, as they were more flexible in their ability to adapt to various shapes and also are very lightweight. Through double layering and the introduction of air pressurization, a degree of thermal resistance was also possible. ETFE could incorporate thin film photovoltaics and frit, to make it more solar responsive. This made ETFE cladding systems and spaceframe systems a good match.

Eden Project in St. Austell, UK, 2001, by Grimshaw Architects used a hexagonal modified spaceframe system to enclose the large domed greenhouses. The outer dome employed a hexagonal geometry created from heavier tube members, which was reinforced with a lighter tubular system as a visually less dominant inner support system layer. Curved triangular truss members formed the junction points between the intersecting domes. The system was able to incorporate operable vents, and its pressurized ETFE façade incorporated varying degrees of transparency to create optimal lighting conditions for the different geographic biomes. The curved geometries were quite regular, which worked with respect to software and fabrication processes available when the project was in development.

A large sloped spaceframe-type structure is being erected on the façade of the Television Cultural Center (TVCC) in Beijing, China, 2012 by OMA. This 3D spaceframe-like all-welded structure uses solid steel nodes and varying sizes of round HSS sections with a somewhat irregular pattern of assembly. The viability of this design choice is in part a function of the acceptability of on-site welding.

Modified spaceframes, those that have morphed into highly customized designs and that are not dependent on proprietary, prefabricated systems tend to be used in regions where extensive on-site welding is more acceptable. This would include Asia and exclude North American and Europe. The idea of the round nodal element, combined with larger, predominantly round HSS sections has been used in several high-profile projects where the geometries have been more challenging and complex, and the loads much heavier than would be addressed in a typical spaceframe.

2. THE EVOLUTION OF THE NODE: DIAGRIDS

Node design is discussed extensively in Diagrid Structures: Systems, Connections, Details in Chapter 6: Node and Member Design.

The function of the node, from its role as a geometric problem solver on spaceframe structures, has evolved to one of accommodating highly complex geometries and stabilizing the steel structure during the construction phase, with minimal need of temporary shoring. Further, on some projects the transfer of forces through the node is critical to ensure the continuing stability of the finished project or building.

Where the nodal connections used in spaceframes were part of a three-dimensional system, often used to carry relatively uniform roof loads or act in a vertical plane as wind bracing, the adoption of the nodal connection type to the diagrid system required significant modifications as a result of the type and size of loads to be carried. The three-dimensional spaceframe was predominantly a spanning application. Diagrids are a single-thickness system, most often oriented vertically and carrying significant gravity loads as well as lateral loads. So although the notion of a central receptor for numerous angular connections remains the same, the loading configuration necessitated a radically different approach to the material and functional nature of this new type of node. Diagrids also represent a significant increase in the structural requirements of lattice systems, which are also single-thickness in nature. Lattice systems can be seen as a contemporized version of more traditional spaceframe systems. These will be discussed in detail in *Chapter 11: Lattice/Gridshell Structures.*

The Swiss Re Building in London, UK, 2004, by Foster + Partners remains a classic example of a pure diagrid tower. Here, the two tubular angled members have been preassembled with the node prior to lifting into place. The assembly must remain stiff to resist deflection to ensure a proper and trouble-free fit.

For tower projects, the node was part of the essential inventive nature of this new means of resisting gravity and lateral loads without the aid of a central core to provide lateral reinforcement. The configuration of the node needed to be substantially different from traditional framed connections given the increase in the number of connecting members and the introduction of non-rectangular geometries. In theory, these nodes are situated in a triangulated system, much like a truss, that places the members in axial loading, thereby allowing the node to act more like a pin connection in terms of the transfer of forces. However, this was not possible, as in order to have construction proceed without need for the temporary shoring of the angled members, the node and connections to it had to ensure that deflections were limited by making the node assembly stiff.

This need for stiffness translated into erection requirements. For lighter assemblies, a node would have its two diagonal members attached on site (as the large V-shaped assemblies, are too large to ship preassembled) and erected as an inverted V. For heavier assemblies all elements would be erected separately and bolted or welded together at height. Here, it was necessary for the diagonal members to be stiffer than required in their finished state, as a means to resist deflection and remain in accurate position to accept the nodal connection.

The node design for the Swiss Re Building clearly demonstrates the axial transfer of forces through the center of the node. Although the structure will be concealed, the approach to the connection of the legs to the node is designed to be tight to keep the cladding profile as small as possible. The horizontal rings are designed as tension members and so their detailing is quite different from the round HSS compression elements.

The nodal element (in the foreground) on the Addition to the Royal Ontario Museum, Toronto, ON, Canada, was designed to be integral with one of the diagonal members that remains continuous through the connection. This decision allowed for the complex fabrication of the plate attachments to be done in the fabrication shop, with the incoming members bolted on site. A larger assembly would have exceeded shipping size. A variety of node conditions can be seen, each with unique fabrication and erection requirements.

Diagrid structural approaches on smaller buildings tend to reflect a higher level of geometric complexity, sometimes also incorporating curvature, in keeping with the abilities to generate the same with contemporary software. The geometric complexities of these projects places increased demands on the design requirements for the nodes, and the lack of repetition within the project as a whole adds greatly to the number of unknowns in the erection process. Where uniform diagrid projects have the benefit of repeated tasks and approaches, chaotic diagrids need to address situations on a one-off basis. Even then, the design of the node, its stiffness and the nature of the connection details to adjoining members will be critical design decisions, aided greatly by contemporary software to assist in the process. On some projects the node may not be fabricated as a distinct and separate element but the functionality of the multi-angle connection ability integrated into one member that will remain continuous through the connection. More chaotic projects will have more parameters to contend with and will look to maximize shipping size and perform some preassembly on site prior to lifting. Larger tower types tend to be simpler to erect, as the assemblies are comparatively uniform. These types of projects are highly reliant on advances in software and the interoperability and high level of communication amongst the team members for their viability and success.

3. NODE FUNCTIONALITY

The spaceframe node acted as a coherent object at the center of the connection, able to transfer forces, accept multiple members, and move the act of connection slightly away from the center. From an erection perspective, this flexibility was helpful. This functionality has been exploited to benefit, as the scale of the nodes used in geometrically complex and diagrid-type structures has increased. Whether the node is fabricated via standard steel fabrication methods or is created through a casting process (addressed later in this chapter), the attachment to the adjacent members is distanced from the important center where the load transfer happens. Distancing the member connections from the center also permits more than one ironworker to work on the connection, speeding up the finishing processes.

Many of the strategies for the effective design of complex steel structures arose from a clear understanding of new connection design requirements and how significantly they varied from traditional steel framing methods. Node-type connections work perfectly with many types of complex geometries and feed well into the aesthetic aspirations of the project. From lattice-type systems to diagrid towers to more chaotic assemblages that must accept and transfer loads from a large number of incoming members, a node connection can be seen as a logical solution.

Whereas normal erection methods are aided by gravity for the positioning of the incoming members, angular and asymmetrical members will fight with gravity and require a highly skilled labor force to best understand how to rig and handle the members to ensure a trouble-free erection process. In spite of

The diagrid structure on the Capital Gate Tower in Abu Dhabi, UAE, designed by RMJM Architects, 2012, is fully exposed. The choice of members and node design was critical to the architectural design. No two nodes are identical here. An intumescent coating is used for fire protection. Concrete filling for additional protection was not possible because of the smaller diameters of the tubes, particularly as they decrease in size over the height of the building.

advances in software, projects with extreme variety in erection requirements will tend to rely on the skill of the lead ironworker to set the crane attachment points and determine the lift chain lengths, as these are not often able to be set ahead of time.

Typical orthogonal connections are fairly stable during the erection process, as most members tend to be symmetrical, balanced, and held in place, at least temporarily, by gravity until the final bolted or welded connections can be completed, that is, if there is not a lot of stress in the connection as a result of its unfinished state. This is not the case for angular geometries, and the node is an essential part of maintaining stiffness during the erection process. As can be seen in the erection process of even the simplest diagrid structure, the in-process structure often includes lengths of diagonal members that are left to cantilever out from the connection point. These will only be stable once the next part of the assembly is erected. This need for temporary stability has added a very new requirement to the connection point. It has to provide adequate strength (moment resistance) to limit movement and deflections as the primary means to avoid what could be excessive requirements for temporary shoring for the cantilevering members.

From the perspective of architectural design, it becomes important to understand the impact that nodes will have on the design, fabrication and erection process. This impact is heightened if the choices is to architecturally expose the structure. If concealing the structure, the "trimness" of the node and member connections will impact the potential bulk of the impact of the structure on the façade and interior design. If fully exposed, the nodes can be exploited for their aesthetics. Diagrids are typically a design choice rather than a structural requirement. The ability to expose the structure is a function of the fire protection strategies. This was discussed at length in *Architecturally Exposed Structural Steel, Chapter 5: Coatings and Protection Systems.*

4. FROM DIAGRID NODES TO MORE WIDESPREAD APPLICATIONS

A structural development can be considered as significant if it transforms design thinking. This can certainly be said of the node system of connection that is central to the concept of the diagrid system. Although not all diagrid structures use the discrete node and connecting member system – some structures have nodes as integral parts of the long diagonal members in order to reduce site-welded connections – it does represent a significant proportion of diagrid structures. This methodology isolates the geometrically articulated nodal pieces, placing the challenge of their exacting requirements in the fabrication shop. This allows for ease of fabrication, tighter dimensional controls and better conditions for welding – all critical to ensuring ease of erection and good fit.

The large structural nodes that were originally developed for diagrid structures have initiated a general transformation of structural design for a wide range of non-diagrid buildings. This transformation has been additionally fueled by advances in BIM technologies and an emergent interoperability from engineering to fabrication software for steel design, detailing and eventual fabrication and erection. "Node" has become a widespread structural term used to describe large, typically steel, connections that accommodate a number of incoming angled members. Nodes are now commonly used to handle the construction of a wide variety of geometries, both regular and irregular. The ensuing surge in the use of eccentric geometries parallels advances in digital modeling and the more direct transfer of information from engineering to fabrication.

The Kobe Port Tower in Kobe, Japan, designed by Nikken Sekkei, 1963, is an early diagrid adopter. The external diagrid (in red) is tied back at its horizontal tension rings to an interior diagrid tower (in white) that supports the access stairs and elevator to the viewing platforms. Its detailing and level of structural complexity responded to the available technology of the time.

The Torre Reforma, Mexico City, Mexico, designed by LBR&A Arquitectos with engineering by Arup, 2016, used an exposed steel diagrid system for the front façade of the building. It is pictured here under construction in 2015. The diagonal steel braces can be seen to support a sun shading system. The diagrid application is quite unusual in that it comprises only the front façade, while the rear two supporting walls of the structure have been constructed of cast-in-place concrete.

A closer view of the steel that is situated outside the line of the curtain wall system. The connections have been visually repressed in this all-welded HSS system. The steel that can be seen on the interior has been fire-protected.

The horizontal tension ring system for the Kobe Port Tower was fabricated from a custom steel plate system, with typical detailing replicated throughout the tower. The on-site connections were all bolted. Similar detailing was used to create the node connections at the crossing of the tubes.

ONE MANHATTAN WEST

NEW YORK CITY, NY, USA, 2019

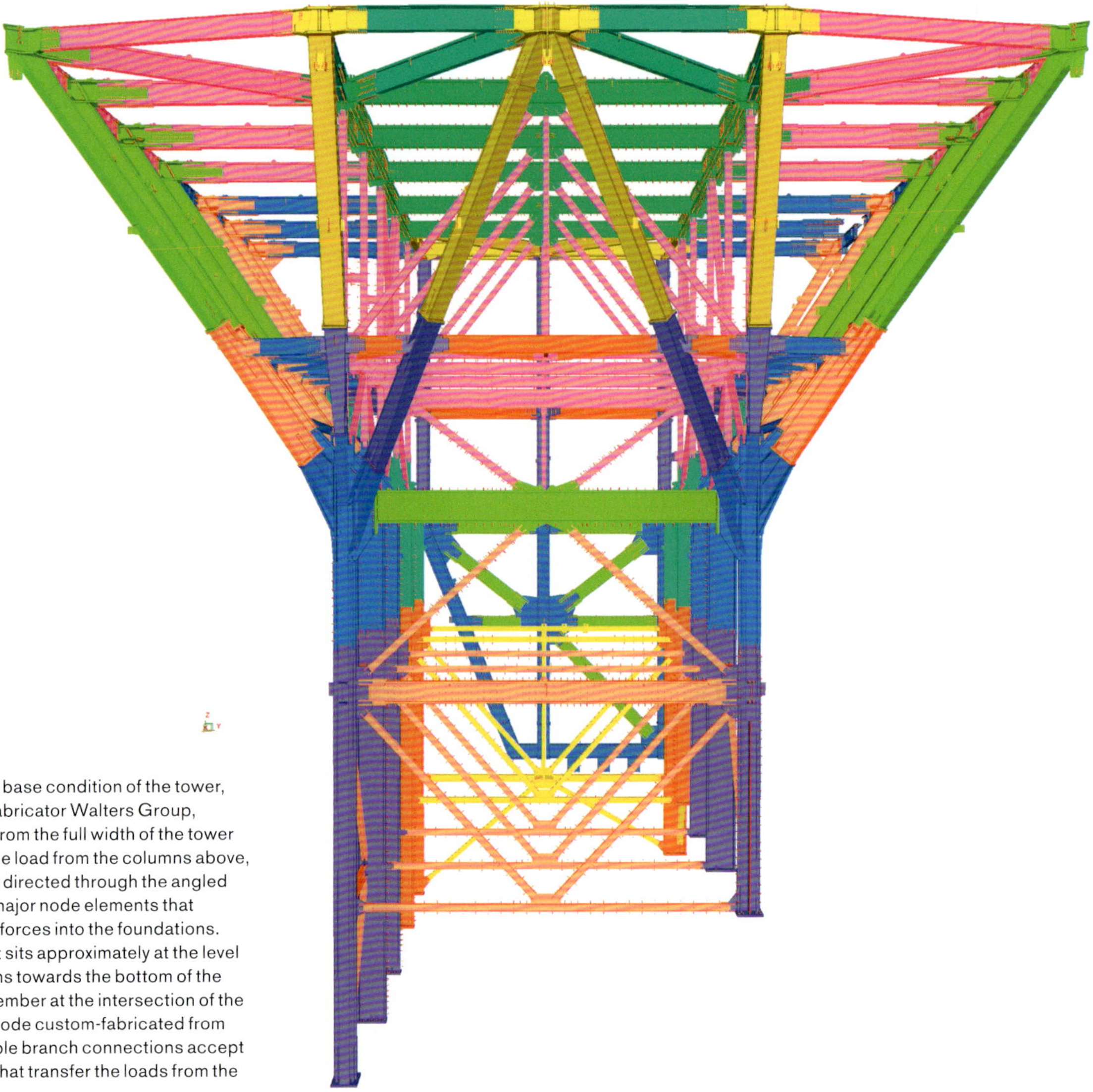

This digital image of the base condition of the tower, produced by the steel fabricator Walters Group, illustrates the tapering from the full width of the tower to its base condition. The load from the columns above, around the perimeter, is directed through the angled members and through major node elements that collect and redirect the forces into the foundations. The grade of the project sits approximately at the level of the light orange beams towards the bottom of the image. The dark blue member at the intersection of the Y is a 51.3-tonne steel node custom-fabricated from heavy steel plate. Multiple branch connections accept the diagonal members that transfer the loads from the perimeter columns.

ARCHITECTS AND ENGINEERS: SKIDMORE, OWINGS & MERRILL LLP

STEEL FABRICATION AND ERECTION OF NODE ELEMENTS: WALTERS GROUP

One Manhattan West is a 67-storey mixed-used tower located in a part of New York City that is undergoing extreme densification. The redevelopment is situated over existing rail lines that were required to be maintained and in service after the completion of construction. This meant that the tower was unable to have a typical foundation system under the perimeter columns and required a radically reduced structural footprint. It was necessary to consolidate the foundations towards the core in order to avoid the underground rail corridor.

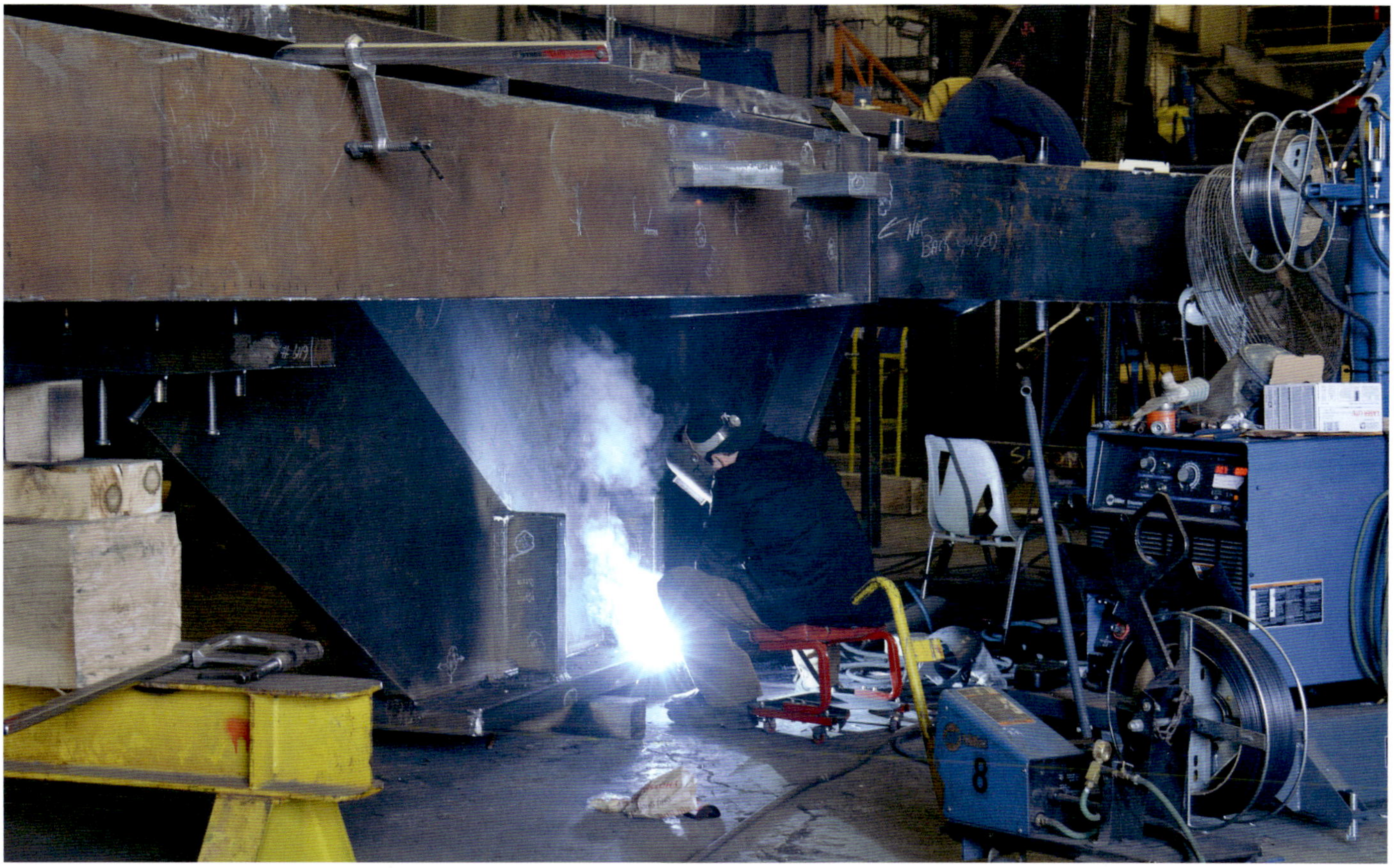

The large steel nodes were prefabricated in the shop of Walters Group in Hamilton, ON, Canada, and shipped by truck to New York City.

Ironworkers at the staging area, preparing the largest node for the lift sequence. The node needs to be lifted to a vertical orientation to be in a good position for its eventual connection to its base. Ideally, the size and tonnage of these types of critical elements allows for the complete prefabrication of the piece in the shop.

The node was provided with a temporary steel base that allows the crane to rotate the element into the vertical position (also aided by a grillage of large timbers that soften the turn) without damaging the column base of the steel node.

One of the logistic issues was that cranes could be located only at the perimeter of the site due to the presence of the underground rail lines. The extreme weight of the node, combined with the reach distance, meant that two cranes were required to complete a tandem lift (see photo p. 55).

The integration of the large nodes and the angular load-transfer components has a ripple effect, requiring complex geometric connections throughout the lower floors of the structure. Bolted connections are used to join incoming members to the shop-fabricated nodal connections, avoiding on-site welding.

What started out as a problem eventually facilitated the removal of traditional column supports from the perimeter of the tower, which cantilevers out over the core. This allowed for the installation of a fully glazed, column-free enclosure system for the lobby.

THE SHED
NEW YORK CITY, NY, USA, 2019

A view of The Shed in its extended position that covers the public square. There is a regularity to the pattern, with heavier vertical members taking the roof loads down to the wheel components. There is also a hierarchy of the diagonal members, with the heavier ones contributing to the load path and lateral bracing, and the lighter ones serving to subdivide the larger diamond shapes and provide better support for the ETFE membrane system. Curvature of the members as they reach the roof level assists in obscuring the roof, which consists of a very robust set of trusses that must maintain the stability between the much lighter side-wall components.

ARCHITECTS:
DILLER SCOFIDIO + RENFRO AND ROCKWELL GROUP

ENGINEERS:
THORNTON TOMASETTI

STEEL FABRICATION:
CIMOLAI

The Shed is an eight-level, 200,000sf/18,580sqm arts center located in the Hudson Yards redevelopment in New York City. It took its inspiration from The Fun Palace by Cedric Price, an imaginative project designed in the 1960s, at a time when technology was not yet capable of incorporating significant dynamic functions. The Shed had the unusual requirement of a movable enclosure. The resultant steel and inflated ETFE-clad structure telescopes over a plaza area to allow for a variety of theatrical spaces, some completely outdoors in good weather.

A view of one of the nodes at the fabrication shop. The bolt-ready end plates allow for discreet connections on site. The access hole on the side of the element allows for the bolting operation. The access hole is positioned towards the conditioned interior side of the wall so that the cover plates need not be sealed against moisture. The elements were primed for corrosion protection prior to shipping.

A view through the interior of the roof component of the movable Shed, showing the use of heavy trusses comprised of wide-flange (Universal) sections. This area needs to support mechanical equipment and is also used to suspend equipment for theatrical productions. The robust design of this element allows the side walls to be lighter both in structure and appearance.

The use of plate steel is essential to the creation of the custom-fabricated curved side faces of the arms. They widen at the top of the side walls to provide increased depth to attach to the roof trusses. Additional connecting elements of varying heights have been welded, at the fabrication shop, to the front face to receive the mounting track for the ETFE system. The curvature on the front face is gentle enough that the self-weight of the steel may be used to ease the material into position for the remaining welding. Where the curvature is more extreme, plate rolling (for even curvature) or brake-forming (for uneven curvature) can be used.

There are precedents for moving roof components, mostly over stadiums, but none to fully move such a large enclosure that was to house a conditioned space. Earlier projects adapted technologies used in the rail industry. The roof of the Skydome Stadium (Rogers Centre) in Toronto, ON, Canada, 1989, one of the first retractable roofs of significant size, covers eight acres, and its weight to be moved was 11,000 tonnes, with a requirement to be fully opened (or closed) in 25 minutes. Naturally, the roof cladding is a significant factor in the dead weight of the structure.

The overall dimensions of The Shed are smaller, but the aesthetic character of the overall form and its material selections are more critical, given the high degree of visibility of the project and its use as a cultural center. The Architecturally Exposed Structural Steel is completely custom-fabricated from curved plate steel and fully remediated where needed to facilitate crisp, curved lines at the meeting points of the plates. While none of the highly articulated steel is actually exposed to view on the exterior, which is either clad in ETFE or metal panels, the diagonal members remain visible on the interior. The kinetic movement of The Shed borrows ideas from overhead cranes, as the structure is also used to lift sets and artwork into the higher-level floors of the adjacent stationary building as well as support performance pieces within The Shed itself when in deployed position.

The custom-designed steel shell is enclosed by a system of ETFE balloon panels. It rolls 114ft/34.7m along two 10in/0.25m-wide, thermite-welded steel rails. Sixteen custom-made steel wheels – each 6ft/1.8m tall – were carefully tapered to glide the structure smoothly into place. The wheels are grouped into six sets (called "bogies"), which carry the 4,000-tonne shell, and driven by six 15-horsepower motors.

Stiffness in the nodal components is essential to limit the need for temporary support during the erection process. Some shoring towers were necessary along the sides of the building to support the steel over the large doorway openings. The truss-like façade was not self-supporting until the majority of the frame was in place and the bolted connections were complete. The arms of the nodal elements could not be allowed to deflect and neither could force be used to coerce the members into place, given the presence of the many cladding attachment pieces.

Given the high level of geometric articulation, great efforts were made to maximize shop fabrication in order to minimize site connections. Here, a test fit of several components is carried out at the fabrication shop in Italy prior to shipping. This is standard practice for complex projects, as a tight fit is required to ensure good alignment.

Inside the member-to-member connection, the bolted connection point is detailed to create a reveal to compensate for any minor alignment issues on site. Access panels over the openings that allow for bolting have been fixed with mechanical fasteners. The gaps have not been remediated, as they will be concealed. The corner welds on the lower portion of the vertical member have used fillet welds, as this portion will be covered. Sharp, fully remediated welds have been used on the twisting curved sections, as they are in view towards the interior space.

The parametrically derived form of the steel for the shed is highly sculptural, with many undulating curves that created predominately diamond-shaped openings. The openings were subdivided into triangular shapes to better accommodate the ETFE skin. ETFE was selected in part to reduce the dead load on the building and also to better accommodate any induced movement in the façade during operation. A glazed façade would have had a higher dead weight and risk of breakage due to movement.

The hollow curved forms of the steel required a high level of customization that included plate rolling, cutting and fully remediated welds to provide sharp edges to all of the corner conditions. The functionality of custom steel nodes was applied to the node-like elements – characterized as such for their ability to assume an unusual, non-orthogonal shape and accept a number of incoming members at varying angles, using non-standard framing details. The larger connecting elements that complete the assembly are fabricated in the same way, resulting in a seamless appearance, as the connection between the node and the incoming members is barely visible due to the continuous geometry and lines of the structure.

In keeping with the common practice in New York City to prefer bolting at height, the connections between the nodes and members use hidden bolts. Unusually, the bolting was carried out on the end-to-end conditions of the members via small access panels that are oriented towards the interior of the building, where they are less visible. This detail is similar to the one used on The Vessel (see p. 89), with engineering and steel fabrication by the same team.

The mounting track for the ETFE membrane system accommodates the curvature and angles of the panels. The detailing is far less refined in this region, as a metal cladding system will be used (see images of the wheels opposite).

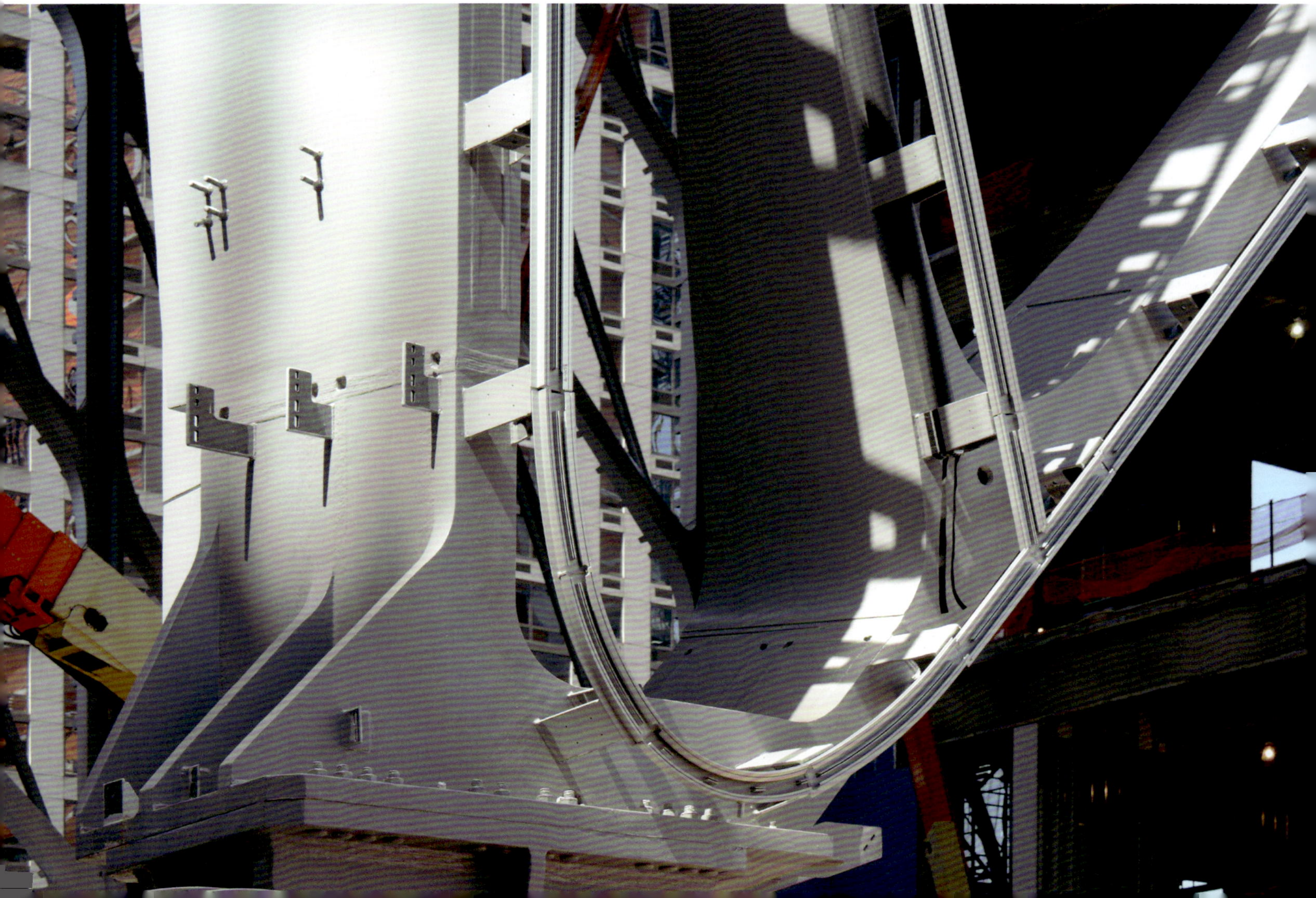

The design and placement of the large wheels was critical to the project. An effort was made in the design to visually soften the technical nature of their presence, yet keep them as a feature of the novelty of the project.

Although a highly technical aspect of the project, the design and finishing of the wheels is featured quite prominently. The cleaning track for the ETFE cladding fits neatly over the seam points and is structurally secured to the steel behind. The nature of this façade, its scale and materiality, exceeds the reach of standard cleaning equipment.

A view up the completed façade showing the corner detail between the overcladding of the structure and the ETFE panels.

A view towards The Shed from The Vessel (see p. 88). The inflated ETFE cladding system gives only a hint at the highly specialized nature of the steel structure behind.

5. CAST NODES

Castings are discussed in great depth in Understanding Steel Design, Chapter 10: Castings.

Two different approaches can be taken to the fabrication of steel nodes to solve complex connection requirements. Nodes can be fabricated using more traditional techniques that employ standard structural sections and plates cut, assembled and welded to the desired shape, or nodes can be fabricated using a method of casting. For most applications, this is a question of aesthetic form in combination with the multiplication factor, node size and budget. If the nodes are to be architecturally exposed, castings can provide a clean solution, allowing the overall form of the structure to dominate the experience, and they are also fairly adept at accommodating geometric variation.

One of the advantages of casting is that the form of the exterior of the node can differ from the shape of the interior void. Depending on the size and required capacity, nodes can be formed hollow or solid. An internal void serves to reduce both the overall weight and the amount of steel required. In cast nodes, steel can be arranged as structurally required, without impacting the desired exterior appearance. It is ultimately the decision of the casting design engineer to address the combination of aesthetic desire and placement of the steel.

Castings require specialized engineering expertise and necessitate a consultant to the primary structural engineer. Forges that can fabricate castings are not yet common. Different foundries use different casting processes, largely based on the size of the casting, its purpose, and the degree of repetition. The most common methods for structural steel components are die-casting and sand-casting. Die-casting is normally used for smaller, repetitive elements such as clevises. Sand-casting is used for a wide range of larger elements, like nodes, that may also be less repetitive. This is often the case in projects with complicated or irregular geometries.

Casting feeds well into current digital practices. A 3D physical model, based upon the digital 3D models created for simulation and design, is used to make the sand-cast. Unique sand-casts can be made for each connector if required. The forms for the castings are referred to as "expendable molds," as the mold is destroyed to remove the steel casting, so it is of less consequence to the casting process itself if the castings are not the same, as the extra expense is largely limited to the preparation of the positive forms to create the outer shapes and the inner voids. The positives used to create the mold can be formed from wood, if reuse is intended, or polystyrene, if a single use is the plan. Extra time and expense comes into play for the engineering required for each unique casting. The resin/sand mixture can be reused for environmental and cost savings.

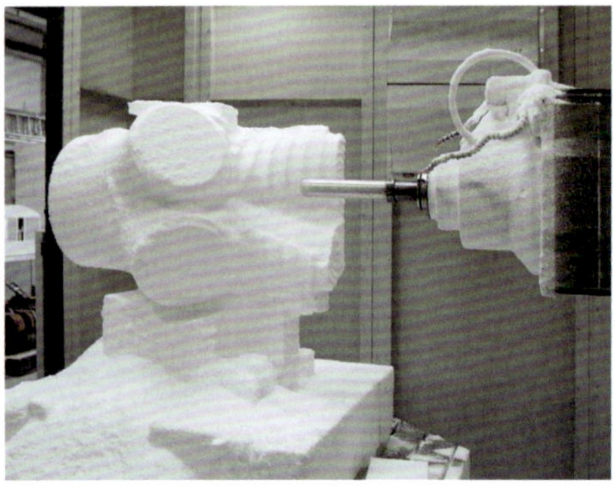

The forms for the positive and the voids of hollow nodes are created via CNC-cutting from the 3D model. A highly accurate form is essential, as any inconsistencies will be directly transferred to the final cast element.

The nodes used at King's Cross Station in London, UK, were cast solid to address high stresses. The deep reveal is part of the node. The connections to the incoming tubular arms and elliptical base are situated at a distance of several centimeters. The purpose of a reveal is to draw your eye away from the fully remediated welded connections.

The node casting at Hauptbahnhof Station in Berlin, Germany, resolves the connection of eight members. The curves have been more easily incorporated into the node, allowing the connecting tubes to be straight. The node has been cast with a step down in diameter prior to receiving the round tubular elements that have had a fully remediated welded connection. If you look very closely you will see faint traces of the remediation on the tubular members. The step detail on the casting is successful at drawing the eyes away from the actual weld location.

Expendable Casting Fabrication

The process for fabricating the types of larger nodal castings appropriate for complex geometries has design implications. Castings have been fabricated that weigh upwards of 22 tonnes. The maximum dimensions and weight is a function of the capacity at the forge (similar to size limitations for galvanizing). The mold assemblies are custom fabrications, but they must still fit inside a protected physical space in the shop and the quantity of molten steel required to fill the mold must not exceed the capacity for a single pour. It is critical that the pour be continuous, as interruptions (cooling) could create unwanted stress points.

Sand-based casting leaves an impression of the granular surface of the sand. Normally, it is necessary to grind the surface of the casting to smoothen out the surface to more closely match the adjacent steel, as hot-rolled carbon steel is typically much smoother. Alternately, it is possible to cast in a slight reveal to draw the eye away from the difference in surface textures. Interestingly, this reveal is normally placed at a slight distance from the actual weld line – perhaps 150 to 200mm away. This permits a more thorough grinding of a smaller portion of the casting and a higher quality of weld remediation.

The casting mold is held in place by a stack of steel frames, with the dimensions and number of stacked frames a function of the size of the element to be cast. The lower half of the mold is first prepared, then the top. If the casting is extremely large, the forms are built in sections and packed into the frame. There is a natural bit of seepage of the steel between the joints of the mold elements, to be ground away during the finish and cleaning process after cooling.

Casting requires non-destructive testing in the fabrication process, as the product could inadvertently include flaws caused by the flow of the molten material through the form or an improperly controlled cooling process. Non-destructive examination should include:

- X-ray examination of first article
- ultrasonic examination of all castings
- magnetic particle examination of all castings
- visual examination of all castings
- chemical and physical material testing of every heat

Both custom fabrication from plate steel and castings provide excellent options in the handling of the joining of complex, non orthogonal geometries.

The forms for the top half of the mold are positioned on top, and spaces are filled with the sand resin mix. Care is taken in the design to understand the flow of the molten steel to ensure that it fills all voids, effectively pushing all air out of the form. Vertical escape tubes allow the steel to properly rise. Horizontal tubes may also be added to facilitate flow between sections. A steel collar holds the forms securely in place.

Another steel collar is placed on top and the remaining voids are packed with sand. Openings are left at the top through which to pour the molten steel as well as select vents to allow air to escape so that there are no trapped voids. The flow path is carefully worked out to ensure that the form is 100% filled.

THE LEAF AT THE DIVERSITY GARDENS
WINNIPEG, MB, CANADA, 2021

The cylindrical diagrid core encloses an internal staircase and serves as the bracing tower for the large cable net system that supports the ETFE membrane.

ARCHITECTS:
ARCHITECTURE 49
AND
KPMB ARCHITECTS

ENGINEERS:
BLACKWELL
ENGINEERING

STEEL FABRICATION:
SUPREME GROUP

CASTING DESIGN:
CASTCONNEX

Designs for greenhouses and botanical garden enclosures have long served to push inventiveness in the combination of steel enclosures and glazing systems. The Leaf at the Diversity Gardens is a 90,000sf/8,360sqm facility that houses a wide range of plant biomes, a butterfly garden and an indoor waterfall. The roof is clad in an ETFE membrane supported on a tensile system, adapted from those used for cable net glazed systems. This system drastically reduces the shadow component of the structural system on the plants.

The premise for the structure is fairly straightforward. A central diagrid core provides vertical access to the multiple levels of the garden experience. Tension cables are stretched from the core to a large rectangular steel ring beam that sits atop the round HSS perimeter columns. The ring is tied down to a series of concrete foundations, one in alignment with each column, to provide the tensioning for the main system of radial cables that support the ETFE roof membrane. Much like a spiderweb, another series of circular cable rings is used to secure the distance between the radial cables. Additional layers of cables are added to ensure a more uniform support system and geometry for the ETFE roof, as the radial geometry results in much narrower panels towards the top of the roof and wider ones around the perimeter.

The curved sections of diagrid are comprised of a series of cast nodes, shop-welded to tubular members to the maximum shippable size. This led to a significant amount of on-site welding to create a seamless look to the AESS feature element.

A tension system around the perimeter of the building allows for the perimeter column system to remain reasonably light in appearance. The glass enclosure system around the perimeter is situated outside the column line.

A close view of the cable net connector system in progress. The stacked method of formed clamping plates allows for multiple levels to accommodate variations in the overlapping geometries generated by the spiderweb-like design.

Castings have thus far not often been used in diagrid towers. Large-scale castings were uncommon during the 2000s and 2010s, when many of the better-known diagrid towers were constructed. Even at that time, digital solutions for structural design and fabrication were in their early stages of development. So much has advanced in the interim, positioning castings as a viable alternative to standard fabrication methods for nodes, with expendable casting methods able to respond to the desired variations in geometry for both the aesthetic drive of the exterior form and the internal void requirements for the strength.

The dense structure of the AESS diagrid support tower in The Leaf, near to view and of limited size, made casting a preferred choice. Casting accommodates the variation in structural requirements of the nodes and keeps the required variations on the inside of the nodes. The Kobe Port Tower, built in 1963 (see p. 136), although significantly higher, is a precedent in terms of its purpose as a viewing tower that encloses an access stair, which is enhanced by views to the expressed steel structure. Although for the 1960s the engineering was quite advanced, the Kobe Port Tower used quite traditional fabrication methods: custom plate nodes were designed to connect straight runs of HSS tubes. The result is a very technical aesthetic.

The Casting Process

The design intent was to create a diagrid tower completely out of round HSS tubes, whereby the intersections could be made without using plate material as connector pieces. The diagrid was to spiral upwards, putting different forces on the tension rings, which normally are in a horizontal position (as can be seen in the Kobe Port Tower). Unlike the visual samples used for many standard-fabricated AESS projects, actual sample castings were not made, as the expense and time required is excessive and a 3D model is able to provide a quick and effective solution.

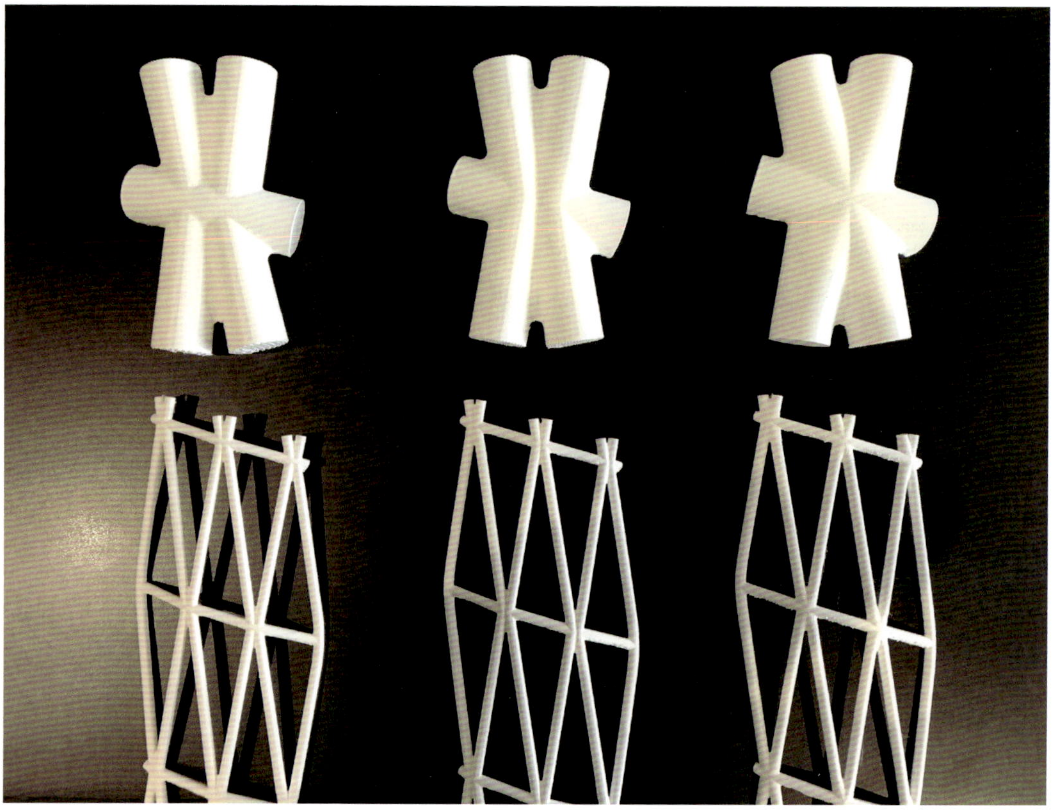

Three versions of the nodal connection have been 3D-printed to look at the aesthetic impact as well as the effect of the geometries on the structural capacity.

The node design was selected to be uniform in appearance throughout the tower, which would also help in creating standardized conditions for welding the connecting tubes. Two strengths were made by changing the wall thicknesses of the node in order to accommodate variations in loading.

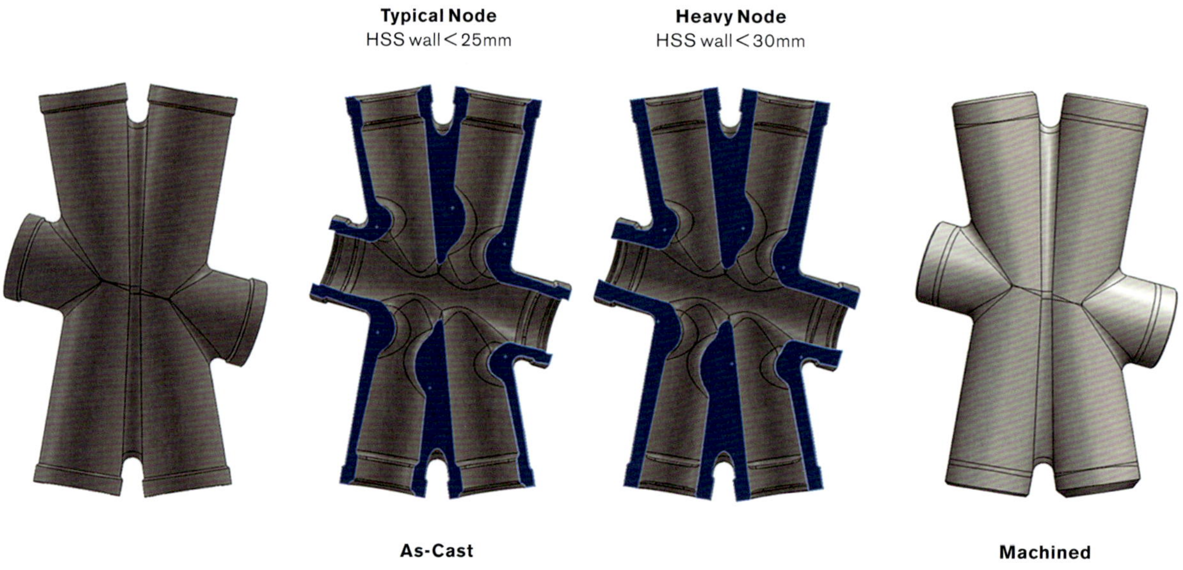

Reusable wooden forms by which to create the sand/resin molds were created using CNC cutting. This allows for a precise and durable shape transfer. The round elements on the face of the mold will support vertical tubes for the pouring of the steel as well as the air vents.

This wooden mold is used to create the sand/resin cast that forms the exterior form of the node. This is called the cope molding. The mold layers are held in place in large steel frames stacked to the depth required for the casting. The layers lock together for security while being moved in the shop and when lowered into the casting pit.

As this node is designed to be hollow, a core mold must also be made that sits at the center of the mold form and creates the central void. The large round ends of the arms extend beyond the node size in order to hold the sides of the two facing molds apart, so that the molten steel can flow freely.

The cast node must be machined so that the ends are properly prepared to accept the connecting HSS tubes. Although the digitally informed process is quite accurate, grinding of the surfaces at the connecting points is required to make sure that the geometries of the node and the tubes align quite precisely.

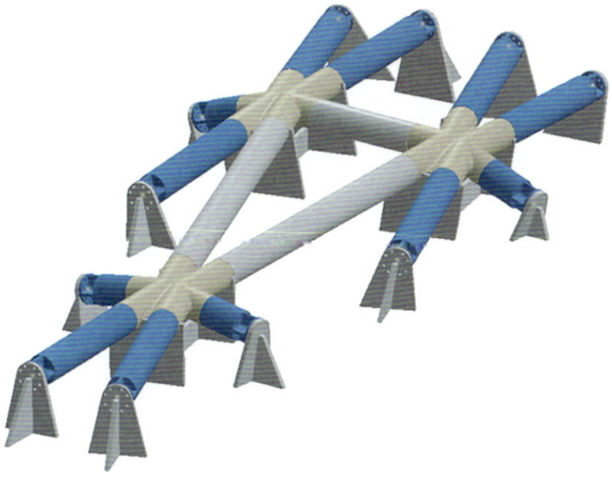

This diagram illustrates the shop setup where three nodes and their connecting tubes are fabricated into a panel ready for shipping. The blue segments give an indication of the extent of the required on-site welding. The original intent was to use a specialized hidden bolted connection by a Diablo connector, but this was abandoned in favor of on-site welding.

QUEEN RICHMOND CENTRE WEST

TORONTO, ON, CANADA, 2016

An overall view of the project showing the position of the two westernmost delta frames. Only the brick shell of the building at the right was maintained. The interior structure was replaced with a reinforced concrete frame system.

ARCHITECTS:
SWEENY&CO
ARCHITECTS

ENGINEERS:
STEPHENSON
ENGINEERING

STEEL FABRICATION
AND ERECTION:
WALTERS GROUP

CASTING DESIGN:
CASTCONNEX

Fairly unusual, the design brief for the Queen Richmond Centre was to erect an 11-storey office tower over two existing historic brick warehouses, thereby creating a large five-storey atrium space at the base of the building. The tower could not bear on the western warehouse, which was to be renovated to join the project, leaving its wooden floors and beams intact. The less beautiful of the two to the east was to have its interior structure removed and replaced with a cast-in-place structure that would better fit in with modern office needs. The decision was made to use a steel system as a support for the reinforced concrete office tower, and turn this into an architecturally exposed element to become the aesthetic signature of the space. Fire protection was a critical aspect of the design: the castings and 1m/39in-diameter custom-fabricated tubular steel legs were filled with concrete and additionally given an intumescent coating.

The decision to use cast nodes as the central connection element was weighed against a more standard fabricated connection that would have employed a cruciform plate system to which the eight tubular legs would be welded. The castings were a premium item but the client agreed that the smoother aesthetic worked better to achieve continuous lines on the massive frames. The castings were designed by Cast-Connex as a consultant to the structural engineer.

Aspects of this project are also discussed in Chapter 5: Fabricating the Steel and Chapter 6: Erection Logistics.

A close view of the Tekla Structures drawing produced by the steel fabricator. Temporary plates were attached to the node and legs to permit the use of bolted erection connections. These stayed substantially in place until the multi-pass welding had been completed. A stepped detail was not desired and so the welding and grinding requirements for the connections were very high. The 3D files are able to be directly used in the fabrication of the molds for the forms, ensuring a precise match between the aesthetic intentions and the final product.

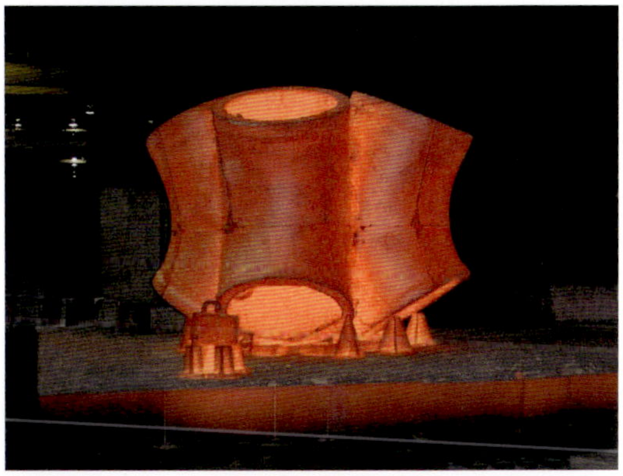

An expendable sand/resin mold, which leaves an imprint on the steel, was used to create the casting. The mold is destroyed in order to remove the casting. The sand/resin mixture can be reused. From this view you can appreciate the capacity of the depression below floor level required to accommodate the mold for the pour. This capacity in part determines the maximum casting size possible at the forge.

The 35,000 pound/15.8 tonne casting must go through a heat treatment process including controlled cooling to calm any interior stresses that may have resulted from uneven cooling, and also to strengthen the steel. Ultrasonic testing is highly recommended for large critical load elements.

The nodes require substantial grinding to ensure that any seam leaks (normal at locations where the sections of the mold meet) are removed. The finish is softened to more closely match the adjacent steel and so that the edges to be welded are accurate in size and properly beveled to facilitate butt welding. The interior void varies as additional steel is positioned to accommodate increased loading, not evident to the viewer. The lugs required for the crane lift are also cast into the element.

A very precise fit is achieved via test fitting between the node and the legs at the fabrication shop prior to the lift. This allowed for a fast removal of the crane. The legs have been primed but the node has not, as the primer would interfere with the grinding of surfaces on the node in preparation for welding.

Wind bracing is provided by an elegant ladder-type truss comprised of a structural HSS central support with stainless steel fittings. The two systems are clearly differentiated in terms of their function through the extreme contrast provided by their materiality and size.

The use of the three large steel delta frames to support the office tower above permitted the use of extensive glazing on the street façades, creating a high level of transparency. This allows for a complementary relationship to the historic brick buildings at the base of the modern tower.

Simple, clean lines were the central focus of the aesthetics that drove the decision to use a cast node at the convergence point of the delta frames for the project. The custom tapered ends of the concrete-filled steel tubular legs required full remediation so that the transitions were seamless.

SALESFORCE (TRANSBAY) TRANSIT CENTER
SAN FRANCISCO, CA, USA, 2018

An overview of the multi-block transit project. Although the perimeter structural system is consistent along the long side elevations, an undulation has been introduced to break up the form. It is achieved through the use of a spaceframe system that ties the aluminum screen back to the primary structure.

The Salesforce Transit Center covers five city blocks in the downtown core, encompassing 1.5 million sf/140,000sqm, and serves as a hub for bus, subway and train connections for the area. To give public space back to the city, the entire roof has been created as a public park. This program asked that the structure be much lighter in tone than would be found in most city-based transit terminals. The multi-storey aspect of the design was also challenged to bring natural light down into the center of the building to illuminate the lower levels. There are three storeys above grade and two storeys below, so illuminating the lower levels required innovation.

A primary structure made of Architecturally Exposed Structural Steel wraps the entire perimeter of the building. It is a seismic-resisting system designed for significant visual exposure. On the interior, the steel is fully exposed, while on the exterior, it sits behind a perforated powder-coated aluminum screen, allowing a partial view to the steel. The aluminum screen is supported on a traditional spaceframe system, which forms a gently curved façade.

ARCHITECTS:
PELLI CLARKE PELLI

ENGINEERS:
THORNTON
TOMASETTI
WITH SCHLAICH
BERGERMANN
PARTNER

STRUCTURAL STEEL:
SKANSKA USA

CASTING DESIGN:
CASTCONNEX

The cast steel nodes for the main structure are crucial elements of both the gravity and the lateral force-resisting systems, as the project is located in a high seismic zone. The design required 304 cast steel nodes, weighing approximately 3.5 million pounds/1,750 tonnes and spread over 75 highly intricate cast node geometries ranging in weight from 6,000 to 45,000 pounds/3 to 22.5 tonnes. The castings belong to two unique systems: those required for the perimeter framing system and those needed for the "Light Column."

Perimeter System

Wrapping the entire perimeter is an AESS frame that comprises the building's Seismic Force Resisting System (SFRS) – an Eccentrically Braced Frame (EBF) arrangement. This is fitted with cast steel nodes at each critical junction: ground level, bus deck level, and roof level. The cast nodes at the ground level integrate 6- to 8-ft/1.8 to 2.4m-deep transfer girders, which span the transverse width of the building. A built-up column sits atop each node. The 32in/0.81m diameter, 2.25in/57mm-thick pipe members are welded to the castings in the field and form the first Y split of the perimeter system. The welds are Complete Joint Penetration (CJP) welds.

The nodes on the bus deck above form the intersection of seven structural members. These nodes are shop-welded (below) and field-welded (above) to 32in/0.81m diameter heavy-walled pipe members. This mechanical pipe is often used in conjunction with cast nodes, as its surface texture is slightly roughened, which matches the node better, and the tubes can be formed without a seam, so that no remediation is needed. The spandrel beams are pin-connected to the sides of the node via 2.25-inch/57mm thick plates that were shop-welded to the node.

The steel castings at the roof level are specially designed Universal Pin Connectors. They have three forks to accommodate the heavy loading that would be transferred in the event of an earthquake. The cast connectors were shop-welded to 32in/0.81m diameter, 1.75in/45mm thick pipe members.

Each tree-like assembly, including the cast nodes, is designed to remain predominantly elastic during a design-level earthquake, with inelastic deformations focused to the shear link portions of the roof-level girder. Given the extremely heavy sections, arduous loading, complex geometry, and architectural significance of the connections, steel castings were a clear choice for the critical junctions of the AESS SFRS framing in this project.

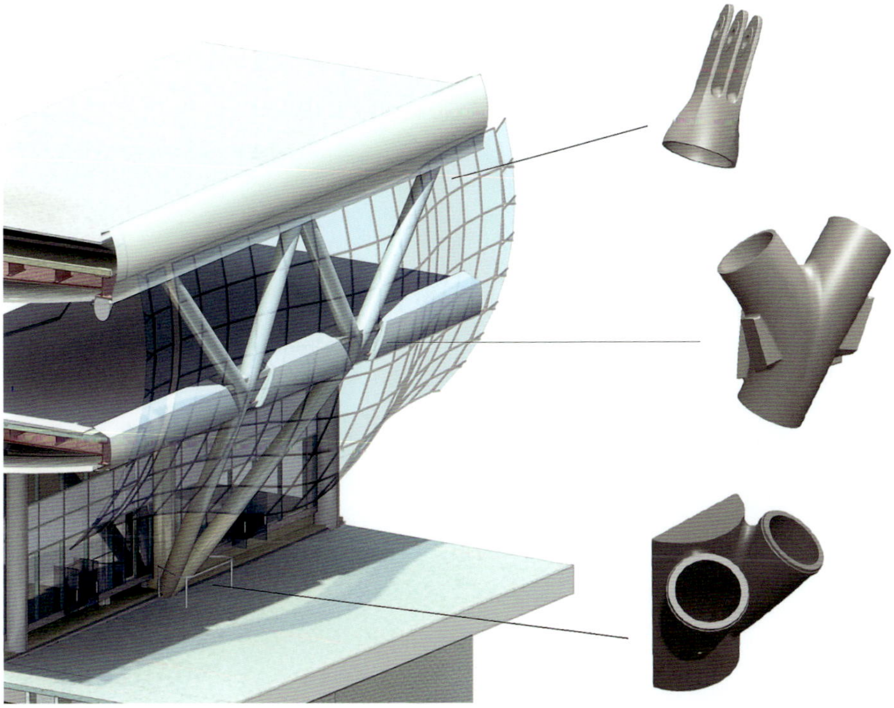

The outward incline of the perimeter framing system was designed to increase the rooftop park area and protect the pedestrian realm below. The castings mediate between large round HSS members. The connections are typically welded with a fairly extensive weld remediation. The perimeter set of castings included typical repeated geometries given the high degree of repetition along the extremely long façades.

The repetitive nature of the multi-pass welded connections allowed for the fabrication of a purpose-built robotic welding device. This saved significant time on the project. The multiplication factor would make this economically viable. The required post -weld grinding, however, was a manual exercise.

Ironworkers are completing the temporary bolted connections prior to the release of the element from the crane. The temporary plates will eventually be removed and the area completely remediated.

Shown here are the pin end connectors that terminate the upper ends of the Y branch columns and pick up the roof beams. Casting in reusable base forms created uniformity in the product, which in the end simplified the nature of the erection.

The corner castings at the ground level required a specialized form to accommodate an additional diagonal member. The temporary plates shown here allow for stability while the connections are prepared and receive extensive multi-pass welding.

The erection included some preassembly on site. Two of the pin-connected diagonals were pre-attached to the upper-level beam section. The required inclines were rigged to the correct angle and held in place by cables during the lift.

Light Column

Cast nodes also feature heavily in the AESS "Light Columns" in the building – one of the main architectural features – that bring light to the lower levels of the building. The 150ft/45.7m-tall light column includes 56 cast nodes with 26 unique geometries. The vertical elements of the light column are composed of gently curving cast nodes between straight pipe segments. The elliptical rings of the light column, except for the top and bottom rings, are made up of roller-bent pipe segments. At the topmost ring, cast nodes are used to transition from the larger-diameter column segments to the smaller-diameter roller-bent pipe segments of the ring. This connection configuration would not have been feasible using conventional fabrication, as the geometry was too tight to use tubular steel between the transitions to the vertical supports, and the required wall thickness too massive to bend.

The top level of the light column will serve as an architectural feature in the rooftop park, so that the AESS aspects were carefully considered, as it is close to view. The scaffold requirements were significant, as the project called for an extensive amount of on-site welding because the overall curved geometry was not conducive to shipping extremely large elements.

Each of the 150ft/45.7m-tall light columns is comprised of 56 cast nodes with 26 unique geometries. As this sectional image shows, the interior geometries of the voids vary to position steel where needed for force resistance and transfer. The lowest-level ring was entirely formed from castings, as the geometry was too tight to use any tubular steel between the transitions to the vertical supports and the required wall thickness too massive to bend.

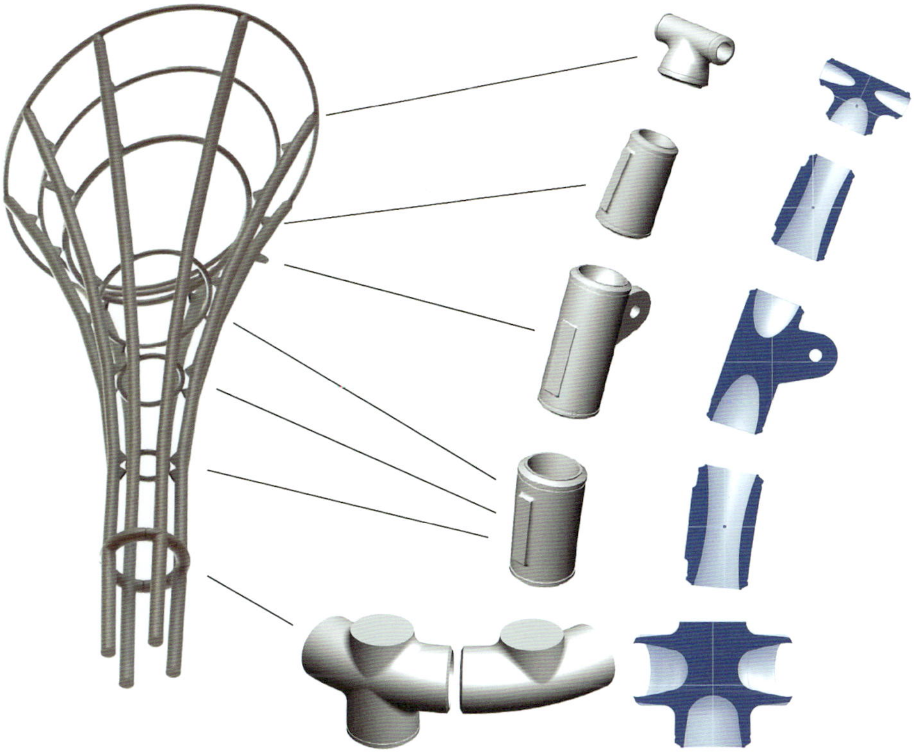

Reveals have been cast into the nodes to draw the eye away from the actual welded connection. Temporary plates and bolts secure the members. The use of 3D models allows for a smooth and accurate shape for the node.

LATTICE/ GRIDSHELL STRUCTURES

11 CHAPTER

1. BASIC DESIGN PARAMETERS

Lattice structures, also called gridshells, are specialized steel framing systems developed to facilitate the creation of complex, curved glazed roofs and façades. Lattice structures work well as a complex system, as their geometry and strength make them ideal for mediating between odd enclosing geometries, and they can also exhibit complexity in form in their own right. They are different from typical skylight systems that frame the glazed panels in a more traditional aluminum frame supported on a system of purlins, beams or trusses. The steel lattice provides the spanning and support for the glazing normally without need for a heavier support structure for the lattice itself. The lattice is designed to be sufficient in strength to achieve the structural span while maintaining a light and modular feel. The structural requirements for the lattice vary due to span, geometry or support characteristics. The strength of the steel is modified by subtle changes in the section properties of the elements rather than by significant changes in the module size that would interrupt the aesthetic. In some instances, slim tensile members are used to provide additional lateral bracing where required. If required, the steel lattice can be fire-protected with an intumescent coating system.

The module for the glazing normally coincides with the module for the steel structure. Therefore, there is direct support for the aluminum framing system of the glazing, which in this way needs not be sized for major span requirements and can be very shallow and visually unobtrusive.

The lattice-supported skylight at Galaxy Soho in Beijing, China, designed by Zaha Hadid, 2012, was a natural choice for providing a simple and clean system that could work with the parametrically designed project. The triangular pattern is easily modified towards the edges to accommodate the tightening of the geometric requirements of the oval shape.

The majority of lattice grids use all-welded connections, as their appearance is much cleaner and the relatively small member sizes may not easily accommodate bolted connections. Generally speaking, as with all complex steel structures, the aim is to fabricate large panels at the shop to minimize on-site welding. As lattice grids are not self-supporting until the system is complete, temporary support for the installation must be provided. When scaffolding is used, no work can proceed under the lattice until it is structurally complete. This has a great impact on the construction schedule.

The major design choices will depend greatly on the desired non-orthogonal geometry of the lattice, the overall span distance and the geometry of the forms between which the lattice mediates. Smaller spans can use more slender members. Tight curvature requires triangulated geometry and also leads to smaller module sizes, to keep the appearance of curvature smoother. Larger module sizes lead to a more faceted appearance to the curved surfaces.

The lattice roof over the courtyard at the DZ Bank in Berlin, Germany, designed by Frank Gehry, 2001, is an early example of the use of a triangulated lattice as a means to accommodate extreme curvature. The barrel vault varies in size along its length. The use of tension members across the space assists in restraining the larger rib elements that control the form. The lattice, covering an area of 19,375sf/1,800sqm, is supported by 2,490 bars and 826 crosspoints (nodes) made of a corrosion- and acid-resistant high-grade steel. There are approximately 1,500 clear glass panels of different sizes. The variation in the glazed panel sizes and angles of the crosspoints was a high challenge given the state of the industry in the late 1990s.

The glazed lattice roof of the Great Court at the British Museum in London, UK, designed by Foster + Partners, 2000, uses a triangulated pattern to mediate the circular geometry of the central gallery and the rectangular shape of the adjacent museum building. The lattice uses an all-welded strategy with visually suppressed nodal connections. The depth of the members, as well as their wall thicknesses, varies with the loading condition. The distance factor makes the connection details and slight structural modifications impossible to view, maintaining an impression of uniformity.

2. MEMBER TYPES

As lattice systems are generally used for skylights, they are located at a distant view and so can allow for some economical approaches to choices in fabrication detailing. The members can use a variety of shapes. Lattices can use HSS sections that may be round or rectangular, simple vertical plates, as well as customized shapes. If standard HSS members can be used, the costs are lower than with custom-fabricated shapes. Many high-profile projects will opt for custom-fabricated sections, as they can more easily be fabricated with a sharpness to the appearance that is not possible to achieve with HSS sections. Also, custom-fabricated structural systems are tuned to the specific requirements of the space, as opposed to aluminum curtain wall framing systems that are mass-fabricated.

A higher degree of custom fabrication of the members is required if there is significant variation in the loading in different areas of the lattice. This is often caused by asymmetrical geometries, as these impact wind and snow loads. At major points of load transfer, for example at supporting columns, the depth of a member is increased to address shearing loads as well as provide stiffness. Variations in loading can also be accommodated by changing the wall thickness of tubular members, thereby maintaining the external geometry, which keeps the nodal connections as well as the overall appearance of the lattice consistent.

3. NODAL CONNECTIONS

Lattice structures normally use node-like connections between the members. The idea of the lattice node is similar to that of the intersection element on a spaceframe system. It is designed to accept often more than four incoming members, at varying non-orthogonal angles, and must be able to maintain geometric flexibility. The key difference between the lattice and a spaceframe, both designed to span long distances in multiple directions in a self-supporting manner, is that a lattice is a single-layer system and spaceframes are spatial structures. The lattice is able to create visual simplicity, as opposed to the visual complexity of a spaceframe. Many spaceframe systems are prefabricated, either proprietary or standardized, and use rather bulky hollow ball-type nodes that allow for a mechanical or a bolted internal connection. By contrast, lattice nodes tend towards an aesthetically driven slimness. Spaceframe nodes are part of the proprietary system, whereas lattice nodes will all be custom-fabricated.

As with all architecturally exposed systems, there is a choice to express or visually suppress the connection between the members. The choice of node type impacts the economy of the system. In the examples that follow, stub sections of round HSS are used to permit the intersection of multiple incoming lattice members with a simple fillet-welded connection method. The use of the stub tube is quite evident even when viewed at a distance. Nodes can also be machined from solid steel, providing flat surfaces on which to weld the incoming members. The choice will be in determining whether or not to allow unremediated fillet-welding, which requires the surface presented by the node to exceed that of the incoming member. Alternately, butt joints can be used, necessitating the sections sizes to carefully match and the welds to be remediated. The latter can be chosen if the intent is for the node and members to blend together, making the act of connection visually recede.

The spaceframe support used at the Baltimore Washington International Airport, USA, is a traditional proprietary system that uses specialized end connectors and ball-type nodes. There is great flexibility at the node to accommodate numerous incoming members. This is a wind support system for a vertical glazed wall that uses a traditional aluminum curtain wall connected back to the nodes as well.

The star-shaped nodal connection for the steel lattice at the British Museum offers a custom fit to the incoming rectangular steel sections. Although there is visual similarity, the nodes are all more or less unique and therefore custom-fabricated. The design visually suppresses the connection.

4. GEOMETRY CHOICES

Advances in computer systems for structural analysis and also for cutting permit the creation of complex lattices that can simultaneously act as structural spanning members and provide direct support for the glass. The geometry of such a lattice is either triangular or orthogonal. Triangulation is more frequently chosen to subdivide the form, as it offers the highest flexibility in accommodating uneven curvature. Triangulation also mediates well when the supporting edges have differing geometries, as demonstrated in the British Museum (see p. 166). Parametrics can be used to modify the pattern during the design phase to find the optimal geometric solution to address concerns related to module, glazing size, steel specifications and erection.

The lattice structure that creates a skylight zone at Tokyo Haneda Airport International Terminal uses double curvature to allow the roof to shed water. Although the skylight appears curved, all of the lattice members have been fabricated from straight sections.

In this structure there is a one-to-one match between the sizing of the glazed roof panels and the lattice. The lattice members are round HSS tubes; the nodal points have been fabricated from short, thick-walled tube sections. The node has been shop-welded to three tube sections. Unremediated fillet welds connect the lattice members to the node, acknowledging the distance factor. Adjoining skylight sections are identified by the presence of an unremediated weld several centimeters away from the hub. A simple clip attaches the skylight frame to the lattice. The connection details are effectively obscured by the lattice when viewed from below.

Both the lattice grid that supports the skylight and the structural steel supporting the glazed retail space at Raffles City, Beijing, China, 2009, use straight steel members to achieve the curvature. This is particularly valuable when the steel frame is intended to support glazed elements, as these typically use a faceted system and flat glass panels. There is a one-to-one coordination between the scale of the lattice and the skylight framing.

The lattice system at Raffles uses T-sections fabricated from plate material as spanning elements between stub-sections of round HSS tubes. A simple fillet-welded connection is used to join the members. The glass is held in an aluminum frame bolted to a plate fin that sits on the top side of the steel framing. The fabrication detailing is not the cleanest but the element sits reasonably out of close view.

The lattice-supported skylight at the mall at KK 100 in Shenzhen, China, designed by TFP, uses a triangulated system to adapt better to the tight areas of curvature at the ends of the space. Different from many lattices, the system uses continuously curved members across the span, triangulated by the addition of straight connecting members between.

The very narrow dimensions of the end condition of the lattice, here viewed as an element in the landscaped roof over the mall, required the use of triangulation to respond to the geometry. The system uses silicon jointing as opposed to traditional mullions to create a smoother aesthetic and aid in water run-off.

The small dimensions of the module, combined with the limited span requirements, allow for the use of uniform depth members fabricated from rectangular HSS sections. Welding visually suppresses the nodal intersection to create a cleaner aesthetic. The attachments for the glazing modules are discreetly hidden on top of the members. This lattice is quite unusual in that the members that span across the short dimension of the skylight have been curved and extend over several modules.

Where the curvature variation is not as extreme, rectangular panels can be used, which may be more in tune with the aesthetics of the space. They will be most applicable when the bounding conditions of the skylight are rectangular. Although triangulated patterns can more easily adapt to a curved geometry, it is possible to make planar adjustments in rectangular patterns, accommodated by adjustments in the glazing attachment system to effect the curvature needed to achieve an aesthetically motivated shape as well as to encourage drainage. It would be advisable to discuss the options with the glazing supplier to understand better how the panel size and geometry limitations need to be accommodated by the design.

The gently curved skylight over the existing courtyard at the Smithsonian Institution in Washington, DC, USA, designed by Foster + Partners, 2007, uses a rectangular lattice. In order to prevent undue loading on the existing exterior masonry wall, the lattice is supported on a series of columns that sit within the space and it cantilevers out to meet and seal at the wall. The undulation of the curved form must be limited so that the glass can remain flat and the slight variations accommodated in the sealing system.

The different levels of stress in the lattice members have been accommodated by a subtle change in depth and width. This allows for uniformity in the spacing, which in turn keeps the glass sizes similar. All of the members have been custom-fabricated to suit their loading. The lattice member is constituted of a round HSS tube forming the bottom edge, with two plate sections forming the sides and a closing plate on top to accept the glazing mullions. The overall cross-section is a hollow trapezoid that is wider at the top flange. The side plates were fillet-welded to the top plate and the round tubular bottom edge. Ribbed panels have been fitted onto the sides of the members, obscuring a view to the welding. Only the bottom edge of the steel lattice is completed as AESS.

The steel lattice that provides cover at the Shiliupu Docks at the Huangpu River in Shanghai, China, engineered by RFR of Paris, 2010, has a rectangular/trapezoidal pattern. This requires detailing the funnel as a series of planar elements. The lattice is entirely comprised of straight steel segments.

Part of the steel lattice canopy at the Shiliupu Docks moves in an undulating curved form, with geometrically careful application of more rectangular shapes. Here as well, the form uses only flat glass panels that require careful coordination to ensure a planar installation.

The members of the lattice are fabricated from steel plates to achieve sharp corners. Welded connections achieve clean lines and downplay the connections. The nodes are simple X-shaped plate elements. The depth of the sections varies with the loading requirements, deepening substantially at the column supports.

A square lattice grid spans between the steel trusses across the column-free Haupbahnhof Station in Berlin, Germany, designed by GMP, 1996. Although the spanning trusses of the large barrel vault are curved elements, the square steel lattice has been constructed using straight segments and the glazing is all flat. As this is an unconditioned space, thermal bridging is not an issue. The clear access to the glazed vault between the external trusses allows for easier cleaning.

The steel sections of the lattice have been fabricated from plate material. To provide rigidity, a doubled system of cross-bracing fabricated from stainless steel has been added to every panel. The top edge of the lattice has been thickened to provide adequate surface area to fasten the glazing system.

5. LARGER LATTICES

A lattice approach can be applied to the fabrication of very large structures. This requires an increase in the size of the spanning members to address increased spans and loading requirements. The strategies at this scale will hybridize the principles of steel lattice grids and those of larger diagrid structures, which combine lateral and gravity systems for larger buildings. Differences in detailing result from their use as a roof versus a wall element, as the detailing for water penetration and drainage is substantially different for the opaque portions of the installation.

The Israel Pavilion at the Shanghai Expo 2010, designed by Haim Dotan, used a large lattice grid to create the structure. With an intended life span of six months, the fabrication, erection and dismantling requirements were different from a permanent structure.

The connections and geometries for all of the glass panels and steel joints were unique. The small scale of the building allowed for greater uniformity in the sizing of the lattice members, making the use of HSS tubes possible. This increased the sizes for discreet bolted connections, which in turn helped site assembly, as it precluded on-site welding. The aluminum glazing mullions have been installed tight to the tubes.

SHAW CENTRE
OTTAWA, ON, CANADA, 2011

The façade of the Shaw Centre is comprised of a steel lattice-framed system, clad in a dark tinted glass to control heat gain into the LEED™ Gold space. A triangulated pattern fits to the irregular curvature.

ARCHITECTS:
BRISBIN BROOK
BEYNON ARCHITECTS

ENGINEERS:
ADJELEIAN ALLEN
RUBELI

The steel lattice of the undulating façade of the Shaw Centre (formerly Ottawa Convention Centre) employs a custom-fabricated system that, although similar in principle to skylight lattice systems, has some unique details. It is used in lieu of a curtain wall, in a vertical position and therefore at a close distance to view. On the other hand, there is no other permanent loading on the steel lattice, as the primary structure of the space behind uses a separate column-and-floor system. The lattice system is designed to handle its own eccentric loading, resulting from its irregular curvature, gravity and wind loading.

A view along the interior of the lattice-supported glazed façade. As shown, the structure for the building is separate and supports the gravity load of the lattice at its bottom edge. The lattice elements have been fabricated from rectangular HSS tubes, with their natural weld seams on the longer sides away from direct view and left unremediated.

The lattice members have been fabricated from rectangular HSS tubes fitted with special end connection elements that allow for an overlapping mating at the custom-fabricated steel nodes. The donut-shaped node is fabricated from solid, machined steel. The incoming elements are tapered in order to accommodate the geometry. The irregular nature of the incoming angles is accommodated by this connection, as it offers rotational flexibility during installation.

A view into the nodal connection, showing how the mating sections overlap. Although all of the nodes appear similar, only some provide for site connections. Larger lattice elements were shop-fabricated to limit site work. The erection required careful sequencing and tracking, as each element is unique and requires erection in a very strategic order.

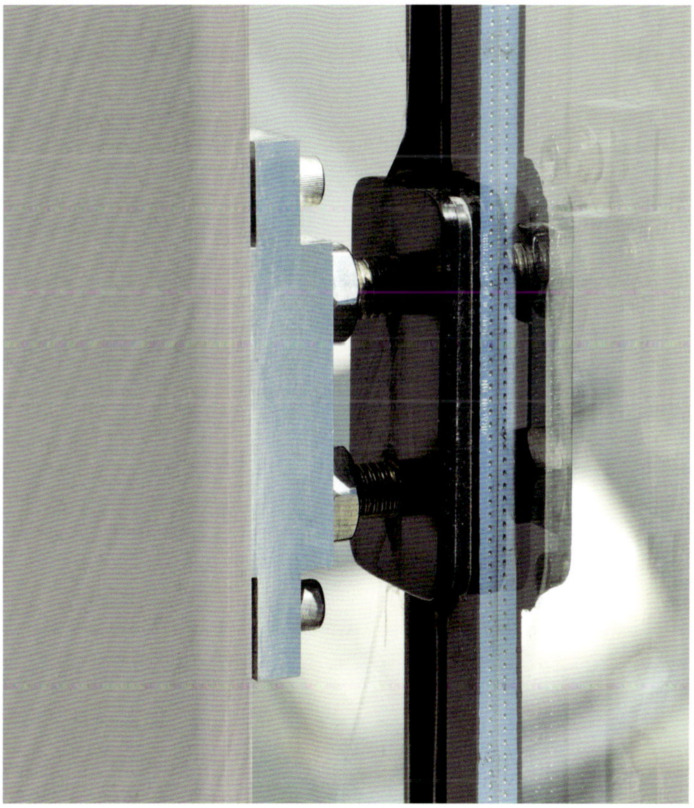

Unlike most lattice systems, this project does not use a standard aluminum framing system to house the insulated glazed panels. Instead, a stainless steel piece is bolted to the rear of the tube. It has a pair of adjustable bolts that connects to individual curtain wall connectors. This allows for the easy exterior installation of the sealed glazed units.

A series of exterior clips hold the triangular glass units in place. The remaining joints are silicone-filled. This has provided a cleaner appearance and better water run-off and precludes difficult fitting of a traditional aluminum system caps at the tight angular nodal points.

KING'S CROSS STATION
LONDON, UK, 2014

Not all steel lattice grids are of the same light scale. Lightweight grids have limitations on span and are best suited to fully glazed applications. When it is desired to also include substantial portions of insulated opaque elements, the loading exceeds the typical steel lattice frame. This was the case for the major Western Concourse addition to King's Cross Station. A new departures hall was to be added adjacent to the historic Western Range portion of the station, whose structure was incapable of providing any support for the roof. The decision was made to use a diagrid structural system, which took a semi-circular shape in plan in order to fit to the curve of an adjacent historic hotel.

Although the overall impression of the 138m/453ft-diameter roof is one of curvature, it was not necessary to fabricate all members to be curved, given the sheer scale of the project. Only the rib members of rectangular section shape have been custom-fabricated as true curves. Welded plate has been used to fabricate box sections with sharp corners. The box-rib radial sections are typically 150mm/6in wide, varying from 250mm-450mm/10in-18in in depth to match the changing bending moments along the arch. HSS sections were deemed unsuitable because of their naturally rounded corners. Plate steel allows for a smooth transition of the changing properties for strength, as they have no natural sectional uniformity like HSS sections. All of the round tubular members that complete the lattice pattern are fabricated from straight sections, affording some economy. The diagrid tubes are standard circular hollow sections, varying between 139mm and 219mm/5.5in and 8.6in in diameter. The wall thickness is also able to vary to suit the strength requirements.

ARCHITECTS:
JOHN MCASLAN AND PARTNERS

ENGINEERS:
ARUP

The cross-sectional properties of the members vary along their length as the funnel opens up to support the roof. The diamond spacing at the base gives way to triangulation in preparation to accept the glazing and roofing components. Triangular geometries are simply more stable.

The erection of the structure required a huge scaffold, filling the entire space beneath the dome, which could completely support the elements as they were lowered into place. The scaffold heights were precisely set to match the curvature, allowing for security and access for welding and bolting operations. Once the shell was complete, the scaffold was sequentially removed in order to allow the now unsupported roof to slowly deflect into its final loaded position. This type of extensive scaffold precludes any work within the building until its removal is complete.

The custom rectangular curved member adjacent to the historic Western Range is much larger than the rest of the radiating curved rectangular steel sections that form the ribs. The diagonal members have been fabricated from round HSS sections, making for easy welded connections to the rectangular members. To accommodate the curvature of the glass roof canopy, small round steel tubes support the aluminum framing system that holds the glazing.

The round tubes that frame into the rectangular rib members use an unremediated all-welded connection. Many of these have been completed on site, given the size of the roof and limitations on the shipping size of prefabricated elements. The distance factor means that these are not readily viewed. The multiplication factor acknowledges the sizeable quantity of these connections and the need for economy. In this location the aluminum skylight framing is closer to the structural steel diagrid lattice than at the entrance canopy, so that the posts connecting the two systems are shorter.

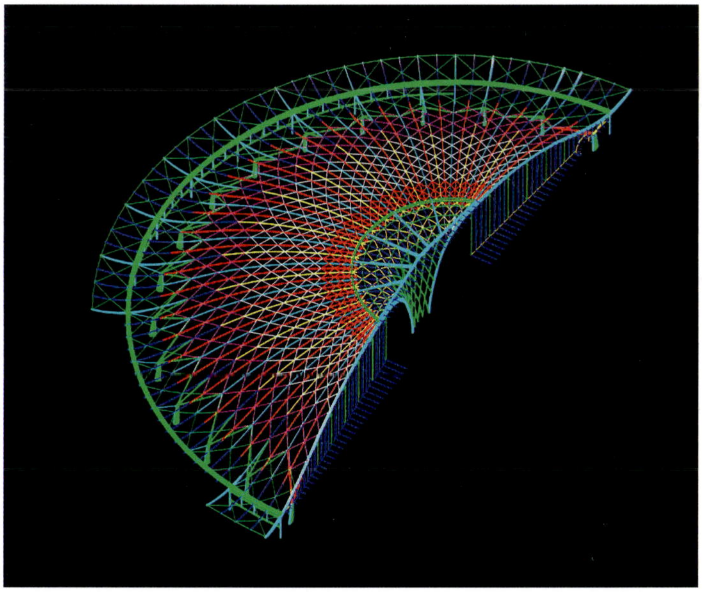

The structural model of the steel structure for King's Cross Station. The diamond-grid lattice uses a funnel shape to concentrate the loads adjacent to the Western Range building. The radiating steel structure rests on a series of columns around the perimeter of the semicircle. There is a natural density to the grid as it rises from the funnel. As the radiating ribs extend to the perimeter, the distance becomes excessive and an additional set of elements is added.

The transition from the top of the funnel base to the spring point of the rectangular arch elements may seem harsh, but is located at a long distance to view. A round plate has been welded to the top of the pair of round HSS tubes to provide a secure, flat surface upon which to weld the roof structure. This is a major point of load transfer. The all-welded solution assists in making the junction less visible from below.

GARDENS BY THE BAY GREENHOUSES

SINGAPORE, 2012

An aerial view of the Gardens by the Bay greenhouses. The complex curved forms required innovative approaches to detailing, as each arch is unique, custom-fabricated from curved segments of plate steel. The glazed envelope is essential for the transmission of light to support the health of the plants.

ARCHITECTS:
GRANT ASSOCIATES
WITH
WILKINSONEYRE

ENGINEER:
ARUP

The large botanical conservatories at Gardens by the Bay use a composite system of large custom-fabricated exterior radial arch supports to suspend the rectangular lattice grid which constitutes their envelope. The radial arches were introduced primarily to transfer the lateral loads to the lattice structure. Positioning these custom-fabricated box arches on the exterior of the environmental enclosure is particular to the hot climate in which the project is situated, where thermal expansion and bridging are not of concern.

A steel lattice system is suspended from the large curved steel arches. Triangulation of the supports to the lattice is used in part to provide lateral stability between the systems. It is more intense at areas where the overall curved form of the lattice grid undergoes a tighter radius. Plate connectors attached to the steel arches and lattice permit a simple clevis connection to be attached at each end of the slender HSS bracing support. This type of pin connection allows for slight modifications on site and responds to the variety of unique geometric connections by allowing for angular variation. Tension cables are used at a high level between the arches for an additional layer of barely visible lateral support.

The concrete foundations are set to receive the thrust loads from the steel arches. The bases use an embellished bolted connection. Plate fins reinforce the welded connection between the arch and the base plate. The arches are made from welded plate steel, with all edges ground and filled to preserve the sharp edges and pristine appearance. Where the radius is tighter, plate-rolling is used to achieve the proper curvature. Straighter sections can often simply allow the self-weight of the steel to naturally deform into the correct curvature.

The intention was to create a large, column-free space that allowed for much variety in planting, to accommodate larger tree forms as well as lower foliage. As the overall goal was to maximize daylight penetration, the members have been shaped to minimize shadows.

The more acutely curved regions of the lattice employ light steel cross-bracing to assist in stabilization. The exterior connections to the large arches are also intensified in these areas.

While the sides of the arch use a simple plate to connect the braces, the rear side employs a cross-shaped plate to simultaneously resist the tension loads and provide stability in the perpendicular direction. The braces connect directly to the steel node of the lattice. The cross-sections of the arch and lattice have been tapered to minimize shadows.

The lattice is fabricated from custom plate formed into a V-shape, providing a narrow edge to the interior and an increased face to the exterior to allow for the attachment of the aluminum glazing frame. The pyramidal nodes are machined from solid steel whose size allows for the use of fillet-welding to attach the incoming members without any need for remediation.

The lattice also has to support operable panels to allow for ventilation. Although glazing does not provide true lateral stability to steel lattices, the addition of operable panes adds complexity to the overall stability of the system and to the requirements for the frame depth and shape.

The operable panels are installed proud of the balance of the glazed envelope and allow water to drain around the individual panels. The glazing uses a silicone jointing system to promote water flow. Though the large steel arches are actually curved, the lattice uses straight segments and flat planes. The sheer scale of the building is able to provide the impression of genuine curvature in its absence.

APPENDIX

SELECTED BIBLIOGRAPHIC REFERENCES

Boake, Terri Meyer. *Understanding Steel Design: An Architectural Design Manual.* Birkhäuser, 2011.

Boake, Terri Meyer. *Diagrid Structures: Systems, Connections, Details.* Birkhäuser, 2013.

Boake, Terri Meyer. *Architecturally Exposed Structural Steel: Specifications, Connections, Details.* Birkhäuser, 2015.

Historical Context

Boles, Daralice D. "The Decon Seven: dismantling a 'movement.' (Museum of Modern Art's exhibit on deconstructivist architecture)." *Progressive Architecture*, August 1988, p. 25+.

Johnson, Philip and Mark Wigley. *Deconstructivist Architecture.* The Museum of Modern Art: Distributed by New York Graphic Society Books, Little Brown and Co. 1988.

Peckham, Andrew. AD Journal Profile 19, Sainsbury Centre, August 1978.

The Deconstructivist Project 25 Years Later. http://www.world-architects.com/pages/insight/deconstructivist-architecture-25

Venturi, Robert. *Complexity and Contradiction in Architecture.* Museum of Modern Art, New York. 1966.

BIM and Digital Tools

Chiang, Lian. "The Software Behind Frank Gehry's Geometrically Complex Architecture." https://priceonomics.com/the-software-behind-frank-gehrys-geometrically/

Eastman, Chuck, Paul Teicholz, Rafael Sacks and Kathleen Liston. *Handbook of BIM: A Guide to Building Information Modeling for Owners, Managers, Designers, Engineers and Contractors.* John Wiley & Sons, 2011.

"Paradigm Shifting: Digital Modeling Mania Upends the Entire Building Team." *Engineering News Record*, June 5, 2006.

Weisberg, David E., "The Engineering Design Revolution: The People, Companies and Computer Systems That Changed Forever the Practice of Engineering." http://www.cadhistory.net/

Bending Steel

Alwood, Todd. "What Engineers Should Know About Bending Steel." *Modern Steel Construction*, May 2006.

"Bending Steel: Questions Answered." *Modern Steel Construction.* June 2002.

Casting Steel

de Oliveira, Carlo, and Tabitha Stine. "Convenient Connections." *Modern Steel Construction*, July 2008.

de Oliveira, J. C., "Steel Castings in Structural Design: Case Studies." *SEAOC Convention Proceedings*, 2015.

"Non-Expendable Casting Techniques." *Castings and Forgings News.* http://www.castings-forgings-news.com/castings/ShowPage.aspx?pageID=1572

Project Profiles

Information for the projects has been obtained through a combination of site visits, discussions with associated firms and text-based sources which are listed below.

BMW Welt Museum, Munich, Germany
http://www.archdaily.com/29664/bmw-welt-coop-himmelblau

Caltrans District 7 Headquarters, Los Angeles, CA, USA
https://www.morphosis.com/architecture/13/

Dalian Conference Center, Dalian, China
http://www.coop-himmelblau.at/architecture/projects/dalian-international-conference-center/

Denver Art Museum, Denver, CO, USA
http://www.archdaily.com/80309/denver-art-museum-daniel-libeskind
https://continuingeducation.bnpmedia.com/courses/navisworks/integrated-bim-and-design-review-for-safer-better-buildings/3/

DZ Bank, Berlin, Germany
https://www.toolcraft.de/en/cases/dz-bank.html

Federation Square, Melbourne, Australia

Bates, Donald L. and Peter Davidson, *Assemblage*, No. 40 (December 1999), pp. 56-67, The MIT Press

Rice, Charles. "Experience and criticality: Returning to Federation Square." *The Journal of Architecture*,Vol. 10, No. 3, 2005.

Gardens by the Bay, Singapore
http://www.dezeen.com/2012/06/19/gardens-by-the-bay-by-grant-associates-and-wilkinson-eyre-architects/
https://www.building.co.uk/focus/gardens-by-the-bay-singapore-/5039760.article

Kings Cross Station, London, UK

The Arup Journal, 2012, Issue No. 2
https://www.theconstructionindex.co.uk/news/view/kings-cross-raises-the-roof

Kurilpa Bridge, Brisbane, Australia

The Arup Journal, 2011, Issue No.1

Lou Ruvo Brain Institute, Las Vegas, NV, USA
http://healthcare.wsp-pb.com/portfolio/lou-ruvo-center/

Louvre Abu Dhabi, Abu Dhabi, UAE
https://www.archdaily.com/883157/louvre-abu-dhabi-atelier-jean-nouvel

Phoenix International Media Center, Beijing, China
http://www.architectmagazine.com/design/buildings/phoenix-international-media-center-designed-by-biad_o
http://www.archdaily.com/165746/in-progress-phoenix-international-media-center-biad-ufo
http://www.biad-ufo.cn/en/projectsshow.aspx?f_id=1908&type=4&years=&workfield=

San Francisco Federal Building, San Francisco, CA, USA
https://www.morphosis.com/architecture/12/

Seattle Central Library, Seattle, WA, USA
http://www.archdaily.com/11651/seattle-central-library-oma-lmn/57219524e58ece408a000001-seattle-central-library-oma-lmn-seattle-in-stability-diagram
http://www.mka.com/projects/featured/seattle-central-library
http://www.architectureweek.com/2007/1003/building_1-2.html

Shenzhen Bao'an International Airport, Shenzhen, China
https://www.archdaily.com/472197/shenzhen-bao-an-international-airport-studio-fuksas

Shenzhen Bay Sports Center, Shenzhen, China
https://www.axscom.co.jp/en/project/overseas/no03341/

Southern Cross Rail Station, Melbourne, Australia
https://grimshaw.global/projects/southern-cross-station/

The Chrysalis, Columbia, MD, USA
https://www.arup.com/projects/the-chrysalis
https://theverymany.com/buildings/13_merriweather-park
https://www.azahner.com/works/the-chrysalis

The Sage Gateshead, Gateshead, UK
http://www.fosterandpartners.com/projects/the-sage-gateshead/

The Shed, New York City, NY, USA
https://www.architecturalrecord.com/articles/14044-the-shed-by-diller-scofidio-renfro-with-rockwell-group
https://archpaper.com/2019/06/the-shed-dsr-rockwell-group-guillotine-doors-facades-plus/#gallery-0-slide-0
https://www.cimolai.com/portfolio/hudson-yards-the-shed/
https://dsrny.com/project/the-shed
https://www.thorntontomasetti.com/project/shed
https://www.archdaily.com/890421/new-renderings-revealed-of-the-shed-at-hudson-yards-as-etfe-cladding-is-installed?ad_medium=gallery

The Vessel, New York City, NY, USA
http://core.thorntontomasetti.com/the-vessel/
https://www.thorntontomasetti.com/project/vessel

Waterloo International Station, London, UK
https://www.dezeen.com/2020/02/27/nicholas-grimshaw-waterloo-station-video/
https://www.youtube.com/watch?v=xJ6dA6f_hRU
https://londonist.com/london/transport/new-platforms-open-at-waterloo-station

Wembley Arch, Wembley National Stadium, London, UK
https://www.designbuild-network.com/projects/wembley/
https://www.brighthubengineering.com/structural-engineering/62918-the-wembley-stadium-arch/
https://www.tekla.com/uk/references/wembley-stadium-arching-ambition
https://www.enr.com/articles/30693-315-meter-span-roof-arch-pivoted-into-place-in-london-6-21-2004

ILLUSTRATION CREDITS

All photographs taken by the author, unless otherwise mentioned.

Cover: The Tekla Structures image of the Canadian Museum for Human Rights, Walters Group

6	The Tekla Structures image of the Canadian Museum for Human Rights, Walters Group
21	The Rooftop Remodeling at Falkestraße, Vienna, Duccio Malagamba
22	The Tekla Structures image of the Brookfield Place Pavilion, Walters Group
26	The Tekla Structures image of the Michael Lee Chin Crystal, Walters Group
37	Ontario College of Art and Design, Walters Group
54	Construction of the Richmond Speed Skating Oval (2 images), George Third & Son
58	The Tekla Structures image of the Puente de Luz, Walters Group
61	Site plan for crane placement of the John Street Bridge, Toronto, Walters Group
63	Fabrication image of the Wembley Arch, Construction Photography, A012-01670, Adrian Greeman
98	The Tekla Structures image of the Brookfield Place Pavilion, Walters Group
101	Plan view of the Sage Gateshead, Foster + Partners
104	Construction image of Southern Rail Station, Melbourne, Scott Sandars
108	The Chrysalis, Marc Fornes, TheVeryMany
109	The Tekla Structures image of The Chrysalis, Walters Group
109	Construction and finished images of The Chrysalis (3 images), Marc Fornes, TheVeryMany
111	Construction images of the Phoenix International Media Center (2 images), BIAD Ufo
118	Interior image of the Dalian International Conference Center, Duccio Malagamba
119	The Tekla Structures image of the AAMI Park Stadium, Arup
122	3D view of the BMW Welt Museum, Coop Himmelb(l)au
123	Construction views of the BMW Welt Museum (2 images), Marcus Buck, BUCK | PHOTODESIGN
125	The Tekla Structures image of the Canadian Museum for Human Rights, Walters Group
126	Drawing showing the evolution of the form of the Tower of Hope, Canadian Museum for Human Rights, Halcrow Yolles
126	3D view of the cloud rails for the Canadian Museum for Human Rights, Halcrow Yolles
126	The Tekla Structures image of a connection for the Canadian Museum for Human Rights, Walters Group
133	Erection of steel for the Swiss Re Tower, London, Foster + Partners
134	Steel node for the Swiss Re Tower, London, Arup
138	The Tekla Structures image of the base of One Manhattan West, Walters Group
142	Construction image of The Shed, Scott Lomax
143	Fabricated element for The Shed, New York City, Cimolai, Scott Lomax
144	Test assembly for The Shed, New York City, Cimolai, Scott Lomax
145	Construction image of The Shed, New York City, Matt Ostrow, Diller Scofidio + Renfro
148	CNC node fabrication, Walters Group
150	Construction image of the Leaf at the Diversity Gardens, Winnipeg, Jacqueline Young, Station Point Photographic
151	Construction image of the Leaf at the Diversity Gardens, Winnipeg (top of page), Bird Construction
152–153	Process images, models and diagrams of the cast nodes for the Leaf at the Diversity Gardens, Winnipeg (7 images), CastConnex
155	The Tekla Structures image of the node connection at the Queen Richmond Centre, Walters Group
155	Production of the cast nodes for the Queen Richmond Centre (3 images), CastConnex
158	Aerial view of the Salesforce (Transbay) Transit Center, Jason O'Rear, Pelli Clark Pelli Architects
159	Diagram of the nodes for the Salesforce (Transbay) Transit Center, CastConnex
160	Erection image of the node for the Salesforce (Transbay) Transit Center, CastConnex
161	Image of the universal pin connectors for the Salesforce (Transbay) Transit Center, CastConnex
161	Erection image of the node and arm assembly for the Salesforce (Transbay) Transit Center, CastConnex
163	Drawing of the construction of the Light Column for the Salesforce (Transbay) Transit Center, CastConnex
177	The Tekla Structures image of the Addition to King's Cross Station, Arup

SUBJECT INDEX

INDEX OF BUILDINGS

INDEX OF PERSONS AND FIRMS

ABOUT THE AUTHOR

Terri Meyer Boake BES, B. Arch, M. Arch, LEED AP

Terri Meyer Boake is a Full Professor at the School of Architecture at the University of Waterloo, Cambridge, Ontario, Canada. She has been teaching in building construction and structures since the early 1980s and launched an effort to expand the environmental core curriculum at the School of Architecture at Waterloo during the mid-1990s. Terri Meyer Boake holds a professional degree in Architecture from the University of Waterloo and a post-professional degree in Architecture from the University of Toronto. She is a LEED Accredited Professional in sustainable building.

She has worked with the Canadian Institute of Steel Construction (CISC), the Association of Collegiate Schools of Architecture (ACSA) and the American Institute of Steel Construction (AISC) to develop resources to promote the teaching of steel construction in schools of architecture across Canada and the United States. She worked with the CISC Architecturally Exposed Structural Steel task force and published the *CISC Guide for Specifying Architecturally Exposed Structural Steel* in 2011. She worked to adapt the CISC AESS Guide for the Australasia market and has toured Australia and New Zealand to promote the system. The American Institute of Steel Construction has also adopted the methodology into their Code of Practice. She lectures extensively on Architecturally Exposed Structural Steel.

Much of this research was used to inform her recent books *Understanding Steel Design: An Architectural Design Manual* and *Diagrid Structures: Systems, Connections, Details* and *Architecturally Exposed Structural Steel: Specifications, Connections, Details*. Her ongoing intense research and investigations into AESS have been essential to the development of this current book, *Complex Steel Structures,* which further investigate issues of steel and non-orthogonal geometries.

She is a member of the Canadian Institute of Steel Construction committee on Education and Research, looking to address the improvement of steel education resources in engineering and architectural programs across Canada.

Terri Meyer Boake is actively engaged with the Council on Tall Buildings and Urban Habitat. She has been a member of the height committee, worked to address the incorporation of a clearer understanding of the nature of composite structures into their buildings database, is a member of the education committee and is a regular contributor to the expansion of the Skyscraper Center database of tall buildings.

She is an avid photographer and has spent significant time traveling to document and bring noteworthy buildings to the attention of the community. Noting a deficiency and lack of critical construction process information available in current publications, her goal has been to document and present buildings with a "construction eye."

The preparation of this book was kindly supported by Walters Group.

WALTERS GROUP

Graphic design, layout and typography:
Reinhard Steger, Maria Martí Vigil
Proxi, Barcelona

Editor for the publisher: Andreas Müller, Berlin
Production: Heike Strempel, Berlin
Paper: 135 g/m^2 Condat Matt Perigord
Printing: Gutenberg Beuys Feindruckerei GmbH, Langenhagen

Bibliographic information published by the German National Library
The German National Library lists this publication in the Deutsche Nationalbibliografie.
Detailed bibliographic data are available on the Internet at http://dnb.dnb.de.

Library of Congress Control Number: 2020941099

This publication is also available as an e-book PDF (ISBN 978-3-03821-430-4)

P.O. Box 44, 4009 Basel, Switzerland
Part of Walter de Gruyter GmbH, Berlin/Boston

Printed on acid-free paper produced from chlorine-free pulp. TCF ∞

Printed in Germany

ISBN 978-3-03821-631-5

9 8 7 6 5 4 3 2 1

www.birkhauser.com